AF183038

One Goal, Two Paths

One Goal, Two Paths

Achieving Universal Access to Modern Energy in East Asia and the Pacific

THE WORLD BANK
Washington, DC

Australian Government
AusAID

© 2011 The International Bank for Reconstruction and Development / The World Bank
1818 H Street NW
Washington DC 20433
Telephone: 202-473-1000
Internet: www.worldbank.org

All rights reserved

1 2 3 4 14 13 12 11

This volume is a product of the staff of the International Bank for Reconstruction and Development / The World Bank. The findings, interpretations, and conclusions expressed in this volume do not necessarily reflect the views of the Executive Directors of The World Bank or the governments they represent.

The World Bank does not guarantee the accuracy of the data included in this work. The boundaries, colors, denominations, and other information shown on any map in this work do not imply any judgement on the part of The World Bank concerning the legal status of any territory or the endorsement or acceptance of such boundaries.

Rights and Permissions

The material in this publication is copyrighted. Copying and/or transmitting portions or all of this work without permission may be a violation of applicable law. The International Bank for Reconstruction and Development / The World Bank encourages dissemination of its work and will normally grant permission to reproduce portions of the work promptly.

For permission to photocopy or reprint any part of this work, please send a request with complete information to the Copyright Clearance Center Inc., 222 Rosewood Drive, Danvers, MA 01923, USA; telephone: 978-750-8400; fax: 978-750-4470; Internet: www.copyright.com.

All other queries on rights and licenses, including subsidiary rights, should be addressed to the Office of the Publisher, The World Bank, 1818 H Street NW, Washington, DC 20433, USA; fax: 202-522-2422; e-mail: pubrights@worldbank.org.

ISBN: 978-0-8213-8837-2
eISBN: 978-0-8213-8870-9
DOI: 10.1596/978-0-8213-8837-2

Library of Congress Cataloging-in-Publication data have been requested.

The views expressed in this publication are those of the authors and not necessarily those of the authors and not necessarily those of the Australian Agency for International Development (AusAid).

Cover photographs
Left: Woman using New Lao Stove. http://www.flickr.com/photos/ashdenawards/4306546092/
Right: Extending the electricity grid to remote areas. iStockphoto.

Cover/book design: LJ Design, Bethesda, Maryland
Typesetting: BMW&W Publishing Services, Baltimore, Maryland

Contents

Box figures

Tables

Foreword

Despite impressive economic growth, in the East Asia and Pacific Region (EAP), over 1 billion people still lack the most basic access to electricity and modern cooking solutions. With approximately 170 million persons lacking access to electricity, EAP lags all other Regions in this respect except South Asia and Sub-Saharan Africa. Almost every second household in the Region lacks access to modern cooking fuels or clean and efficient cooking stoves. This lack of access to modern forms of energy has a direct bearing on achieving the United Nations' Millennium Development Goals (MDGs), which are intended to reduce poverty while increasing education, empowering women, and improving child and maternal health by 2015.

Indoor air pollution from the widespread use of coal and wood-based biomass fuels is responsible for acute respiratory illnesses and related ailments that lead to over 600,000 premature deaths in EAP every year. Women and children are especially vulnerable to this exposure which, in the case of particulate matter in indoor smoke, can reach up to 20 times the safety levels recommended by the World Health Organization (WHO). Large populations in the EAP countries will continue to depend on coal and biomass fuels for several decades. Thus, the most pragmatic approach to reduce indoor pollution and its harmful effects is to move toward advanced cooking stoves. So far, most EAP countries have made only small and scattered efforts to promote advanced cooking stoves, and even these efforts have been constrained by institutional and financial shortcomings.

Meeting the enormous challenge of providing for universal access to electricity and modern cooking fuels and advanced cooking stoves in EAP requires the governments of EAP countries to work simultaneously on two paths. First, achieving universal electricity access requires accelerating both grid and off-grid programs while employing appropriate policies and innovative technical solutions to reduce costs, improve reliability, and provide timely service to all EAP households. On the second path, a major push is needed to increase access to clean cooking fuels (natural gas, liquefied petroleum gas, and biogas) and advanced cooking stoves, particularly those utilizing biomass in poor rural areas. If the goal of universal access to electricity and modern cooking solutions is to be achieved by 2030, efforts to promote them must be scaled up massively.

Both of these paths are affordable. The combined investments required for a scenario of "Universal Access" to electricity, modern cooking fuels, and advanced cooking stoves are estimated at US$78 billion over the next 2 decades. This amount represents an increase of US$32 billion over the amount required to maintain the "Business-as-Usual" scenario over the same period. The annual incremental investment needed for universal access is only 0.1 percent of the Regional GDP, excluding China. Nevertheless, if most of the poorer EAP countries are to reach the goal of universal access to modern energy by 2030, they will require significant support from donors and multilateral institutions.

The World Bank is well positioned to assist medium- and low-access countries to accelerate their electrification programs and open a second front in fighting energy poverty by increasing access to modern cooking solutions. In doing so, the Bank can build on its global knowledge and experience in promoting access to modern energy, including successful rural electrification programs in the Region. In consultation with the EAP countries and other development partners, the Bank also will work toward establishing a Regional forum that would share information and promote activities to catalyze access to modern energy in the Region. This facility would focus on the poorest and most remote populations of the countries, who otherwise might not gain access to modern energy in the next two decades.

James W. Adams
Vice President
East Asia and the Pacific Region
The World Bank Group

Acknowledgments

This flagship report was undertaken by the Infrastructure Unit of the Department for Sustainable Development in the East Asia and Pacific (EAP) Region of the World Bank under the guidance of John Roome, Sector Director, and Vijay Jagannathan, Sector Manager.

The energy sector leader for the report was Dejan Ostojic. The core team comprised Ramachandra Jammi, Douglas Barnes, Arun Sanghvi, Subodh Mathur, Yabei Zhang, and Laurent Durix.

The project team is grateful to a number of World Bank colleagues who acted as advisors and peer reviewers. They include Sameer Akbar, Pedro Antmann, Veasna Bun, Enrique Crousillat, Gailius Draugelis, Koffi Ekouevi, Julia Fraser, Defne Gencer, Mohinder Gulati, Wendy Hughes, Ede Ijjasz-Vasquez, Migara Jayawardena, Anil Malhotra, Dana Rysankova, Yanqin Song, Richard Spencer, Jie Tang, Natsuko Toba, Alan Townsend, Hung Van Tien, and Jia Zhenhang. The team also wishes to acknowledge Raffaella Centurelli, Energy Analyst, International Energy Agency; Alan Coulthart, Lead Specialist, AusAID; and Anthony Jude, Director, Energy, South-East Asia Department, Asian Development Bank, for their contributions as external peer reviewers.

The team benefited greatly from a wide range of consultations in Cambodia, China, Indonesia, and the Lao People's Democratic Republic (PDR). The team wishes to thank the participants in these consultations, who included government officials, nongovernmental organization and civil society representatives, the private sector, and donors.

The editorial and production team was led by Alicia Hetzner. Patricia Katayama, Acquisitions Editor, Office of the Publisher, was always available for consultation.

The team would like to acknowledge the continued generous support from the Government of Australia, which funds the EAP Energy Flagship Report series. This series includes *Winds of Change: East Asia's Sustainable Energy Future* (2010) and the present volume, *One Goal, Two Paths: Achieving Universal Access to Modern Energy in East Asia and the Pacific*.

Abbreviations

AAA	Analytical and advisory services
AC	Alternating current
ADB	Asian Development Bank
ADO	Automotive diesel oil
AECOM	(Global firm)
AEI	Africa Electricity Initiative
AMS	Approved small-scale methodology
ASTAE	Asia Sustainable and Alternative Energy Program
BioCF	BioCarbon Fund
BoP	"Bottom of the pyramid" (model)
CDM	Clean Development Mechanism
CFL	Compact fluorescent lamp
c/kWh	US cents per kilowatt hour
CO	Carbon monoxide
CRESIP	Cambodia Rural Electrification Strategy and Implementation Plan
DC	Direct current
DC/AC	Direct current/alternating current
EA	Energy Authority (Mongolia)
EAC	Electricity Authority of Cambodia
EAP	East Asia and the Pacific Region (World Bank)
EC	Electricity cooperative
EdC	Electricité du Cambodge
EdL	Electricité du Laos
EPIRA	Electric Power Industry Reform Act (Philippines)
ERI	Energy Resources Institute
ESCO	Energy service company
ESMAP	Energy Sector Management Assistance Program (World Bank)
ETS	Emissions trading system
EVN	Electricity of Vietnam
FEA	Fiji Electricity Authority
FIP	Forest Investment Program
FJc	Fiji cents
FSM	Federated States of Micronesia
GDP	Gross domestic product
GEF	Global Environment Facility
GHG	Greenhouse gas
GIS	Geographic information system
GIZ	Gesellschaft für Internationale Zusammenarbeit (formerly GTZ)
GNI	Gross national income
GOI	Government of Indonesia
GOP	Government of Philippines
GOT	Government of Tonga
GPOBA	Global Partnership on Output-Based Aid
GR	Government Regulation
HH	Household(s)

HIV/AIDS	Acquired Immunodeficiency Syndrome/Human Immunodeficiency Virus
HV	High voltage
IBRD	International Bank for Reconstruction and Development (World Bank)
IEA	International Energy Agency
IFC	International Finance Corporation
IHP	Indoor household pollution
IIT	Indian Institute of Technology
IPP	Independent power producer
Km	Kilometer
KW	Kilowatt
kWh	Kilowatt hours
Lao PDR	Lao Peoples' Democratic Republic
LDU	Local distribution unit
LED	Light-emitting diode
LGU	Local government unit
LPG	Liquefied petroleum gas
LV	Low voltage
MDG	Millennium Development Goal
MEDP	Missionary Electrification Development Plan (Philippines)
MEM	Ministry of Energy and Mines (Lao PDR)
MIME	Ministry of Industry, Mines and Energy (Cambodia)
MMRE	Ministry of Mineral Resources and Energy (Mongolia)
M/T	Metric ton
MV	Medium voltage; megavolt
MW	Megawatt
MWSS	Manila Water and Sewerage Services
NaS	Sodium-sulfur (battery)
NGO	Nongovernmental organization
NISP	National Improved Stove Project (China)
NLS	New Lao Stove
NPC	National Power Corporation (Philippines)
NTT	Nusa Tenggara Timur Province (Indonesia)
O&M	Operations and maintenance
ORE	Office of Rural Electrification (Thailand)
PC	Power Corporation
PDR	Lao People's Democratic Republic
PEA	Provincial Electricity Authority (Thailand)
PERTAMINA	Perusahaan Tambang Minyak Negara (National Oil Co.) (Indonesia)
PICs	Pacific Island Countries
PLN	Perusahaan Listrik Negara (State Electricity Company) (Indonesia)
PM	Particulate matter
PNG	Papua New Guinea
PNPM	National Program for Community Empowerment (Government of Indonesia pilot)
PoA	Programme of Activities (CDM)
PPA	Pacific Power Association; power purchase agreement
PPL	PNG Power Ltd
PPP	Public-private partnership
PRIF	Pacific Region Infrastructure Facility
PROPER	Program for Pollution Control, Evaluation, and Rating (Indonesia)
PSO	Public service obligation
P2P	Power to the Poor (Lao PDR)
PV	Photovoltaic

QTP	Qualified third party
R&D	Research and development
RE	Rural electrification; renewable energy
REAP	Renewable Energy for Rural Access Project (Mongolia)
REC	Rural energy company (Cambodia)
REC	Rural electric cooperative (Philippines)
REDD	Reducing Emissions from Deforestation and Forest Degradation
REE	Rural Electricity Enterprise (Cambodia)
REF	Rural Electrification Fund
REN21	Renewable Energy Policy Network for the 21st Century
RMB	Renminbi
RMI	Republic of Marshall Islands
SFC	Specific fuel consumption
SFU	Solid fuel
SHS	Solar home system
SME	Small and medium enterprise
SOE	State-owned enterprise
SPC	Secretariat of the Pacific Community
SPUG	Small Power Utilities Group (Philippines)
SREP	Scaling up Renewable Energy Program
Sq km	Square kilometer
SSC	Sales and service center
SWAp	Sector-wide approach
SWER	Single-wire earth return
TA	Technical assistance
TERM	Tonga Energy Roadmap
TPL	Tonga Power Ltd.
UC-ME	Universal charge for missionary electrification (Philippines)
UN	United Nations
UN-DESA	United Nations Department of Economic and Social Affairs
UNDP	United Nations Development Programme
UNEP	United Nations Environment Programme
VER	Voluntary emissions reduction
WB	World Bank
WBG	World Bank Group
WESM	Wholesale electricity market trading
WHO	World Health Organization
WSP	Water and Sanitation Program (World Bank)
WTS	Wind turbine system

Key Messages

Achieving universal access to modern energy is within the reach of countries in the East Asia and the Pacific (EAP) Region in the next two decades. Some EAP countries have practically achieved universal electricity access. Others that lag behind recognize electrification as a major governmental priority. Nevertheless, the Region is still far from achieving universal access to modern energy given that more than 1 billion people—or every second household (HH) in the Region—lack modern cooking solutions. Indoor pollution from solid fuels using traditional and inefficient cooking methods is a leading cause of health problems for women and children in the Region, and a major barrier to achieving the Millennium Development Goals (MDGs). However, with the right policies and approaches, universal access to energy is within the reach of the Region in the next two decades.

Improving access to modern energy requires the governments to work concurrently on two paths. On the first path, achieving universal electricity access requires accelerating both grid and off-grid programs through applying appropriate policies and innovative technical solutions. These solutions can reduce the cost, improve the reliability, and provide timely access to all EAP HH by 2030. On the second path, a major breakthrough is needed to increase access to modern cooking fuels (natural gas, liquefied petroleum gas, and biogas) and advanced (clean and efficient) cooking stoves, particularly those utilizing biomass in poor rural areas.

Both paths are affordable but require significant financial support in low-access countries. The combined investment requirements for universal access to electricity and modern cooking facilities are estimated at US$78 billion over the next 2 decades. This estimate represents an increase of US$32 billion over the "Business-as-Usual" scenario until 2030. The annual incremental investment needs would be a small share (0.1 percent) of the Regional GDP (excluding China). Nevertheless, poor, low-access countries would require significant concessional financing of the order of one percent of their GDP annually.

The World Bank Group is committed to expand policy dialogue, knowledge-sharing, and financing to help EAP countries achieve universal access to modern energy. The Bank has accumulated global knowledge and experience in promoting access to modern energy, including successful rural electrification programs in the Region. Thus, the Bank is well positioned to assist medium- and low-access countries to accelerate their electrification programs and to open a second front in fighting energy poverty by increasing access to modern cooking fuels and advanced cooking stoves. In consultation with client countries and its development partners, the Bank also will initiate the establishment of a Regional forum that will promote access to modern energy in rural areas in EAP countries, particularly those that are not likely to obtain access to grid electricity or modern cooking options in the next two decades.

Executive Summary

Two Faces of Energy Poverty: Lack of Access to Electricity and Modern Cooking Solutions

Energy poverty still affects a large share of the population in East Asia and the Pacific. In the past three decades, the East Asia and the Pacific (EAP) Region has experienced high economic growth and rapid urbanization. In parallel, energy consumption has more than tripled and is expected to further double over the next two decades. Nevertheless, approximately 170 million people (or 34 million households) in EAP countries do not have electricity connections in their homes. This number is equivalent to approximately 9 percent of the Region's population, and 30 percent of the Region's population excluding China. Approximately 1 billion people (or nearly half of all households in the Region) still use solid fuels including coal and wood-based biomass for cooking, primarily with traditional and inefficient stoves. The lack of access to electricity affects primarily rural areas. In contrast, the lack of access to modern cooking solutions (modern cooking fuels[1] and advanced cookstoves[2]) is a broader phenomenon that affects both rural and urban areas in EAP countries (figure 1).

Access to modern forms of energy has a direct bearing on the achievement of the United Nations' Millennium Development Goals (MDGs). The MDGs seek to reduce poverty while increasing education, empowering women, and improving child and maternal health to agreed levels by 2015. Access to modern energy is a crucial input to meeting the MDGs, and requires progress on both paths: electricity and modern cooking solutions.

The benefits of increased access to electricity are high for poor people. The welfare benefits of rural electrification for a household (HH) adopting electricity typically range from US\$10–US\$20 a month, or up to US\$1 per kilowatt hour (kWh). For household lighting alone, the benefits are

1. In this report, "modern cooking fuels" refers to natural gas, liquefied petroleum gas (LPG), and biogas. "Traditional cooking fuels" and "solid fuels" refer to wood-based biomass fuels, agricultural residues, and dung. For cooking, "modern energy" and "traditional energy" are used to distinguish between ways of using energy rather than a type of fuel. Thus, the use of natural gas, LPG, and biogas, as well as solid fuels in efficient or less polluting stoves, is considered a set of modern cooking solutions. The use of solid fuels in inefficient or open stoves is considered a traditional cooking method. Coal, charcoal, and kerosene are seen as "transition fuels," which are best used with efficient or less polluting stoves.

2. Recently, "advanced cooking stoves" has come to represent a newer generation of stoves that have higher combustion efficiency and are manufactured in either workshops or factories. These stoves can represent major progress over the earlier generation of "improved cooking stoves."

Figure 1. Rural-Urban Divide in Access to Modern Energy in EAP

1(a) Population without access to electricity, 2008 *(%)*

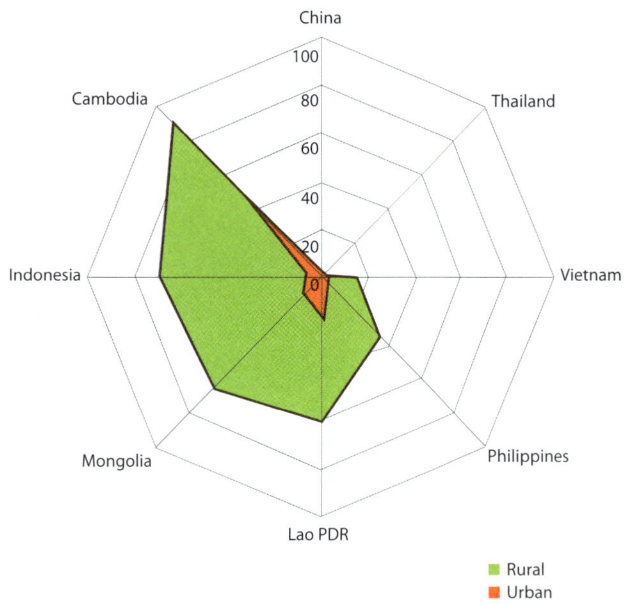

Sources: IEA 2009; authors' calculations.

1(b) Population not using modern cooking fuels, 2007 *(%)*

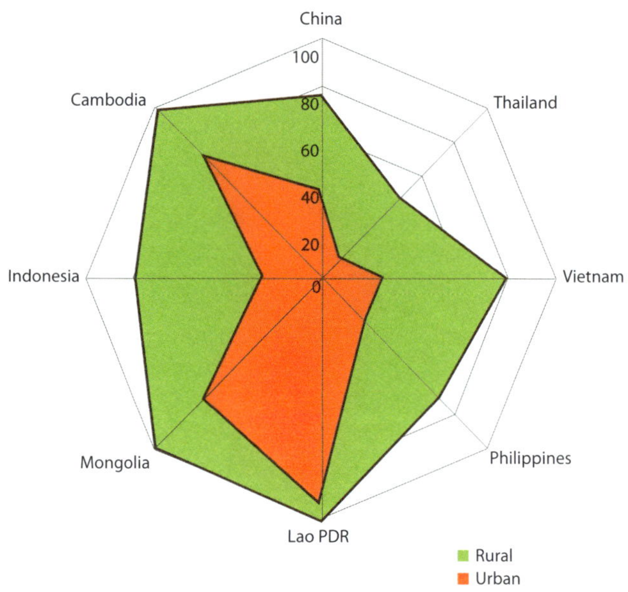

Sources: WHO and UNDP 2009; authors' calculations.

estimated to be close to US$0.80 per kWh in Lao PDR and US$0.50 per kWh in the Philippines. These benefits are much higher than the cost of supplying electricity to rural areas, which ranges between US$0.15–$0.65/kWh.

Improved access to modern cooking solutions can go a long way in improving health and reducing premature mortality, especially among women and children. In HH who rely significantly on biomass for cooking using traditional methods, women and children are exposed to air pollution levels in the form of small particulates from smoke that can reach 20 times the maximum recommended levels. Children and women also are more likely to suffer the drudgery of gathering wood and other biomass. Indoor smoke pollution from inefficient use of biomass for cooking is estimated to cause over 600,000 premature deaths annually in EAP. In the absence of a suitable policy framework that sustains market-based solutions, these numbers are expected to rise in keeping with trends in developing countries worldwide. This trend contrasts with other leading causes of premature deaths (HIV/AIDS and malaria), which, given the continued support of active programs, are expected to decline (figure 2).

Universal access to modern energy is within the reach of EAP countries, but they must make significant progress on both paths: electricity and modern cooking solutions. While many EAP countries have made good progress in providing electricity, among all regions, EAP has the largest number of people who do not use, or have access to, modern cooking solutions. In urban areas, the challenge is to provide modern cooking fuels and advanced (clean and efficient) cookstoves to the large numbers of HH who lack them, and electricity to the fast growing numbers of HH. In rural areas, the challenge is to provide both electricity and modern cooking fuels and advanced stoves to the vast majority of people, including those who live in remote, isolated, and/or sparsely populated areas.

Barriers to improving energy access include a lack of awareness and commitment as well as institutional and financial constraints. A major barrier

Figure 2. Lack of Access to Modern Cooking Fuels and Its Relative Health Impacts in Developing Countries, 2008–30

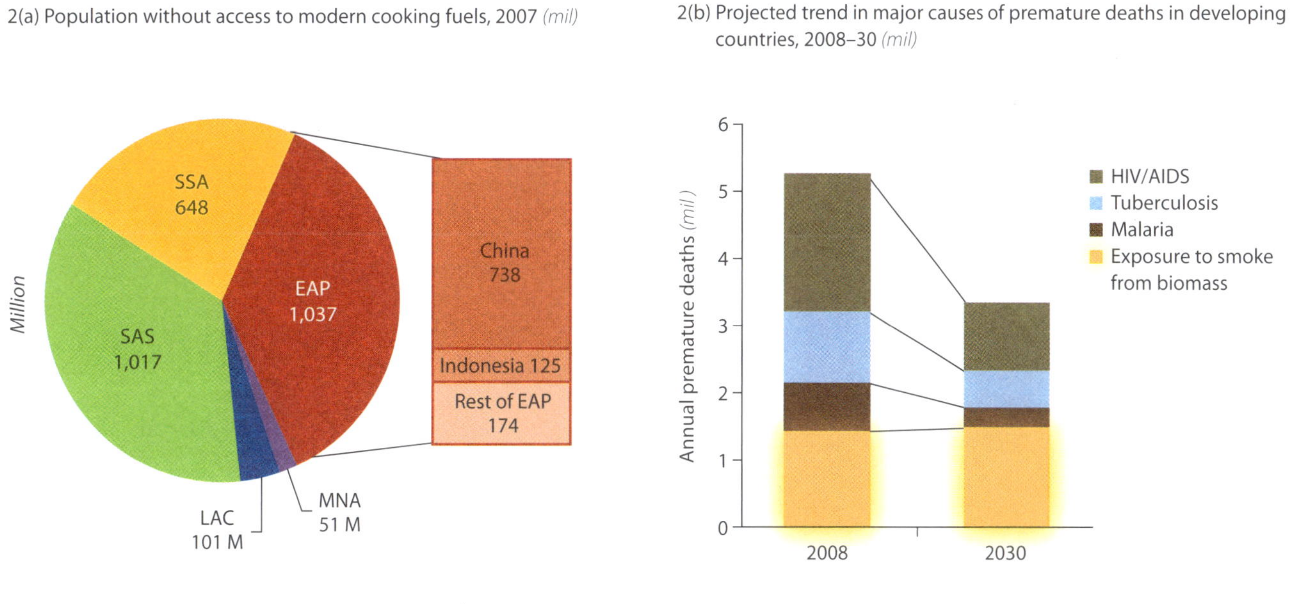

Source: WHO and UNDP 2009.

Source: IEA 2010.

to addressing energy access issues is a lack of awareness among policymakers as well as potential beneficiaries—especially regarding modern cooking solutions—and inadequate attention to the problem at the higher levels of government. Provided that government commitment is forthcoming, the main barriers to universal electricity access are limited institutional and implementation capacities, and low levels of affordability on part of the beneficiaries. Regarding modern cooking solutions, a major additional barrier is the lack of low-cost, reliable LPG supply and clean and advanced cookstoves that poor people can afford. Fortunately, there is adequate experience worldwide and within EAP to form the basis of the policy interventions needed to achieve the goal of universal access to energy.

Electrification and Development: Fighting Poverty and Stimulating the Economy

The last decade has seen significant progress in increasing electricity access in several EAP countries, but the overall picture is still quite diverse (table 1; figure 3). China, Thailand, and Vietnam are close to universal electricity access with more than 95 percent of HH having been electrified. Cambodia, Indonesia, the Pacific Island Countries (PICs), and the Philippines continue to have large numbers of unelectrified HH and, more worrisome, have not had high rates of increase in access in the last decade. Interestingly, Lao PDR, one of the EAP countries with the lowest GDP per capita, has achieved the highest electrification growth rate in the Region in the last decade. However, the country faces the challenge of maintaining this rate as it now must serve increasingly remote and dispersed HH. Mongolia's main challenge is to provide cost-effective electricity access to its nomadic people.

Table 2 describes the main challenges that EAP countries face in increasing their electricity access through grid and off-grid solutions.

Two Access Scenarios for Electricity Access: Business-as-Usual and Universal Access

This report has developed two scenarios for electricity access in EAP up to the year 2030: (1) a "Business-as-Usual" scenario, based on current policies and trends in EAP countries; and (2) a Universal Access scenario, in which all EAP HH would have access to electricity by 2030

Table 1. Electricity Access in EAP, 2009

Country	Electricity access (%)	Population without electricity (mil)
Indonesia	65	81.4
Myanmar	13	43.9
Philippines	84	15.0
China	99	8.0
Cambodia	24	11.4
PNG	7	6.3
Vietnam	96	3.6
Lao PDR	70	1.9
Timor-Leste	22	0.9
Thailand	99	0.7
Mongolia	90	0.3

Source: IEA 2010; authors' estimates.

(figure 4). These scenarios are not forecasts or plans. Their purpose is to help policymakers in EAP countries and their development partners explore policy options and appreciate the incremental financing needs associated with achieving universal access to electricity.

Both scenarios in figure 4 take into account all of the available technical options to provide electricity access: main grid connections in urban and rural areas; rural minigrids; and rural HH systems, such as solar photovoltaic (PV) systems. Both scenarios, including the roles and shares of

Figure 3. Growth in Electricity Access versus GDP per Capita in EAP, 2000–09

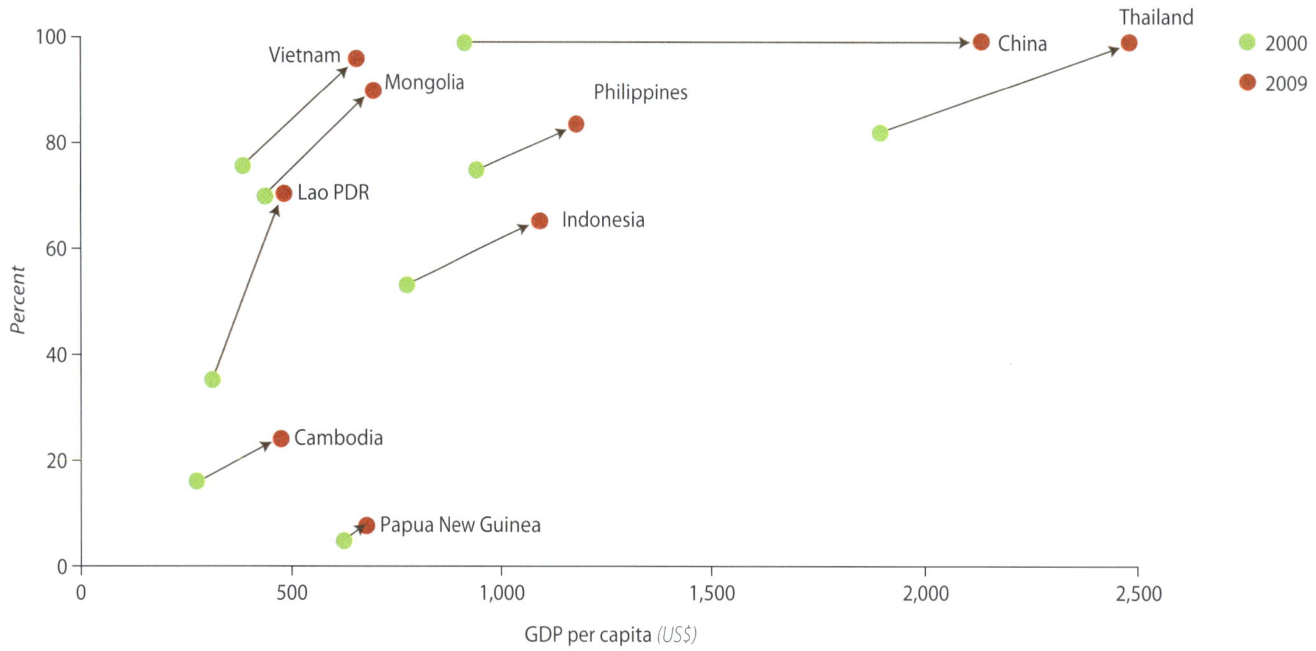

Sources: IEA 2010; World Bank 2010e; authors' calculations.

Table 2. Electricity Access Challenges in EAP

Level of electricity access *(% HH)*	**Grid**	**Off-grid**
High access *(>95)* China, Thailand, Vietnam	Finalizing "last-mile" issues	Innovating energy solutions for remote HH
Medium access *(50–95)* Indonesia, Lao PDR, Mongolia, Philippines	Maintaining momentum of programs; jumpstarting programs that have stagnated	Solidifying existing efforts and making necessary reforms to serve communities and HH in remote areas
Low access *(<50)* Cambodia, Myanmar, most Pacific Island countries (PICs), and Timor-Leste	Getting started and making a serious commitment to expand national grid	Developing the institutional and regulatory framework for off-grid solutions

Source: Authors.

Figure 4. Electricity Access in EAP: Business-as-Usual and Universal Access Scenarios, 2010–30

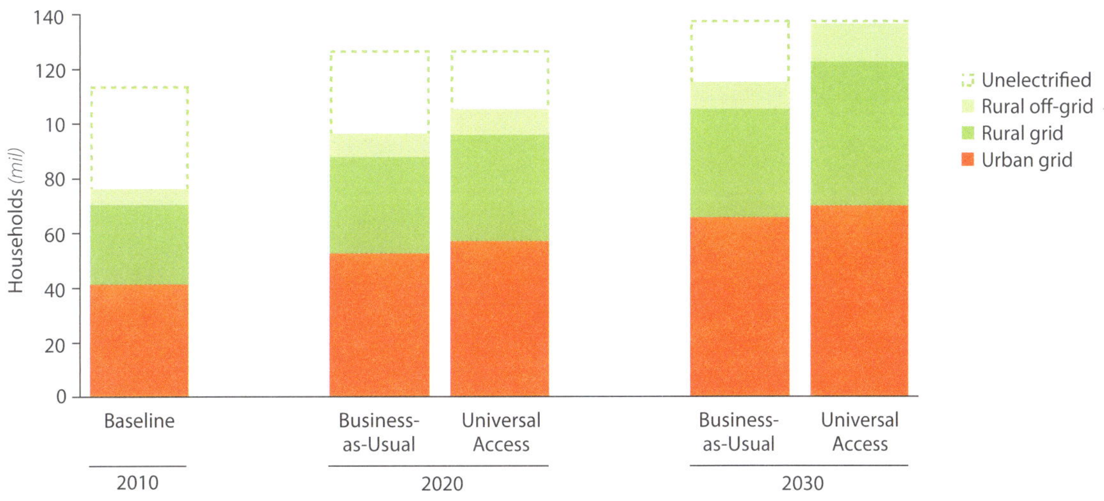

Sources: IEA 2010; UN-DESA 2008; authors' calculations.

these technical options, are based on country-level analyses that take into account their specific challenges (table 2) and projected trends in population growth and rural-urban migration.

Under the Business-as-Usual scenario, the number of HH without electricity access would decline from approximately 40 million in 2010 to 24.5 million by 2030. While nearly 40 million HH would be connected over 2010–30, population growth (at an average rate of 0.45 percent per year) would create new HH and increase the need for electrification. In 2030 approximately 9 million HH in Indonesia still would lack access to electricity—the largest number in any country in the Region. The focus of the Business-as-Usual scenario is grid-based electrification, particularly in urban areas, even though the number of HH electrified by off-grid options also would increase from 6.5 million in 2010 to approximately 11 million in 2030.

Figure 5. Schematic Profiles of Relative Costs of Electricity Supply: Main Grid versus HH Solar PV Systems

5(a) Cost curve for HH solar PV service in villages remote from main grid

5(b) Cost curve for HH solar PV service in villages 3–5km from main grid

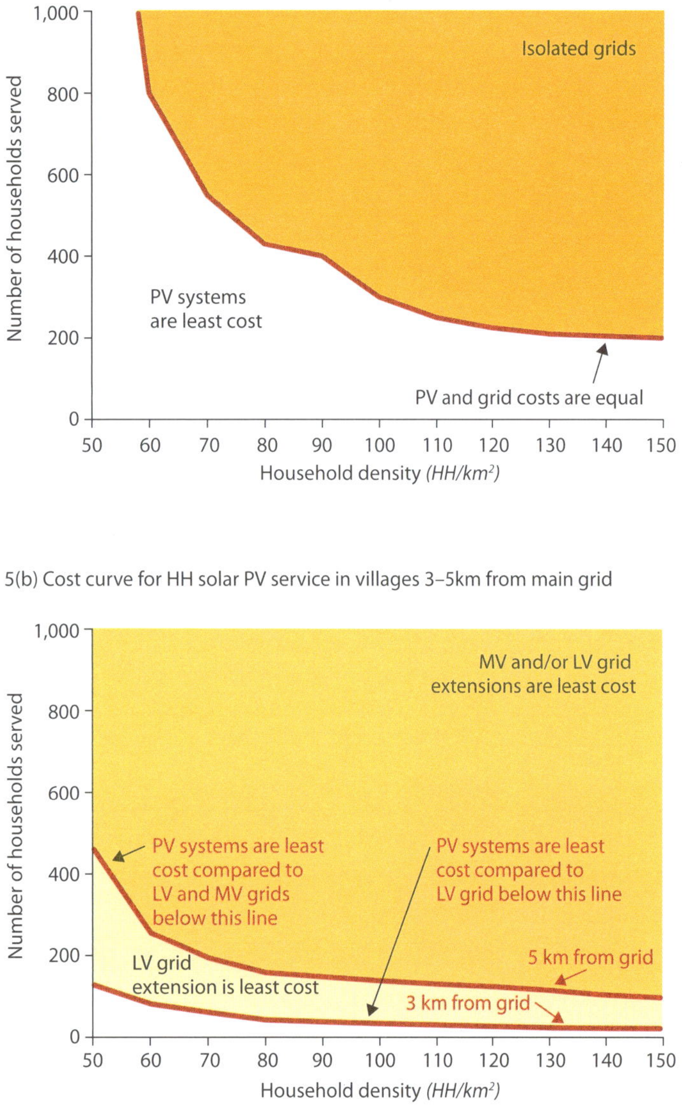

Sources: World Bank 2008d ; authors' calculations.

The Universal Access scenario additionally intensifies the focus on minigrid and HH energy systems. Although, in the Universal Access scenario, grid-based electrification would deliver the largest increase in electricity access, the use of off-grid and HH energy systems would be comparatively greater. Without this greater use, it would not be possible to serve remote HH in a cost-effective and timely manner. EAP countries that have achieved near-universal access, such as China, Thailand, and Vietnam, already use minigrid and HH energy options for remote and sparsely populated areas.

Grid and off-grid options are cost-effective for different population densities and geographic segments. Grid systems are least cost for serving large loads in concentrated areas. Isolated grids are least cost for serving smaller, localized loads in places far from the grid. Household energy systems, such as solar PV, are least cost for low load areas that are far from the grid. Figure 5 shows schematic cost-effectiveness boundaries among grid electricity service, minigrids, and solar PV service, based on HH density and distance of the village from the main grid.

Both scenarios require subsidies to maintain (Business-as-Usual scenario) or to accelerate (Universal Access scenario) the rate of rural electrification. All rural electrification worldwide and in EAP has been based on some form of subsidy. While it is considered best practice to provide subsidies for capital cost only, subsidies for variable cost commonly are employed for reasons of equity and can be justified if they are efficiently designed, properly targeted, and effectively implemented. The need for subsidies would remain even after steps have been taken to reduce costs. For example, Lao PDR has introduced many cost-reducing innovations in its network. Doing so has reduced, but not eliminated, the need for subsidies.

The level of capital subsidy is a key criterion for defining the technology mix that maximizes the electrification rate while meeting service standards. Since subsidies are needed, the overall fiscal affordability of the capital subsidy requirements plays a key role in selecting technical options for rural electrification. For example, a country that is close to universal access, and has ample resources to provide subsidies, may choose grid extension to cover its unserved households with the same quality of service as the rest of the population.[3] However, in EAP, it has been common to use lower-subsidy, lower-service off-grid and HH energy systems as an option to maintain the fiscal affordability of the subsidies.

The reliability and quality of electricity supply are important factors in delivering the benefits of electrification. If there are grid power shortages or network deficiencies, as is common in some EAP countries, rural consumers may suffer blackouts and brownouts (low voltages), particularly during periods of peak demand. This unpredictability limits the potential benefits of electrification and discourages productive activities and even connecting to the grid. In contrast, rural HH solar PV systems may not suffer from these problems. However, these systems usually are designed to provide only a limited amount of power for a few hours a day. Therefore, meeting acceptable service standards is an important planning criterion in defining the optimal technology mix for rural electrification.

Timely electrification is another critical criterion, if universal access is to be achieved by 2030. Consider a situation in which a minigrid is cheaper than grid extension—but the implementation capacity of the established grid operator is much higher than that of a potential minigrid operator. In this case, it may make sense to proceed with grid extension, provided that the grid is backed by adequate generation capacity and the incremental subsidy requirements are fiscally affordable.[4] In some situations, the grid operator may find it difficult to provide service to remote communities. In these cases, HH energy systems may provide a timely solution until these areas are scheduled for grid extension in future periods, based on the country's grid extension plans.

3. For example, Brazil has extended the grid to isolated and remote communities in the Amazon, for which costs are extremely high (Niez 2010).
4. This has been the approach in South Africa (Niez 2010).

The above criteria—reliability, quality and timeliness—are interrelated. Furthermore, to some extent, these criteria themselves depend on government actions and policies. For instance, reliability and quality depend on the actions taken to augment the capacities of the grid operator, potential minigrid operators, and HH system installers. Reliability and quality of service provided by the grid also depend on actions taken to add and maintain adequate generation capacity and the network. The costs of the grid and minigrid options—which have implications for timeliness—depend on the introduction of cost-reducing innovations and practices. In addition, the costs of HH energy systems may depend on cost-saving arrangements such as bulk procurements. In practice, at different times, the role of the various technology options would be based on the country-specific current and expected future values of costs, capacity, and availability of subsidy funds.

Policy Interventions to Achieve Universal Access to Electricity

Accountability for results is more important than the specific organizational structure for electrification. Experience shows that achieving a high level of electricity access requires a competent and responsive national institution accountable for delivering results on the ground in terms of reliability, quality, and timeliness. In many cases, this role has been performed by power utilities, which also serve as strategic reservoirs of technical "know-how" for planning, design, implementing, and operationalizing services required to scale up off-grid and HH energy systems. However, as stressed above, the specific form of the organization is less important than the principle of accountability—a core value adopted by the most successful electrification programs.

The best off-grid electrification programs combine dedicated institutional and financial support with public-private partnerships. The experience from various off-grid electrification programs (most of them outside EAP) indicates the importance of two success factors: a specialized off-grid agency and a targeted off-grid funding mechanism. At the same time, a dedicated government agency is needed to create a stable regulatory environment and set technical standards, but without prescribing specific technologies. The latter should be left to the entities implementing off-grid electrification. The specialized off-grid agency can be well positioned to administer subsidies from the off-grid electrification fund to the implementing agents, including private, nonprofit, or nongovernmental organizations (NGOs). Using such a public-private partnership (PPP) model can add the synergistic benefit of promoting small and medium-sized enterprises (SMEs) in parallel with the off-grid electrification program.

A system of tariffs and subsidies should ensure sustainable cost recovery while minimizing price distortions. The high costs of electricity supply in rural areas and the limited capacity of households to pay for the service make it difficult to attract investment in rural electrification. The tariff and subsidy mechanism should supplement the revenues that the utilities receive from consumers with subsidy funds to match costs that would be incurred by an efficiently run service provider. This complementary revenue should be paid upon confirmation of delivery of service of adequate quality. This subsidy would ensure that the utilities remain focused on rural service provision. Subsidies could be funded through contributions from the government as well as from within the power sector. The latter could be made through cross-subsidies within a consumer category (such as lifeline tariffs that favor households with low consumption), between consumer categories (as from large industries to residential), and/or from urban to rural consumers.

Improving environmental sustainability and reducing the cost of off-grid electricity are possible through a transition to renewable energy, but service levels may not always be comparable to the grid and conventional diesel minigrids. Most minigrid systems still are based on diesel generation, but often are too small to take advantage of economies of scale. Furthermore, the availability of diesel fuel in remote locations often is expensive and uncertain due to transportation and storage difficulties. On the other hand, renewable systems, such as micro- and minihydro, wind-based systems, hybrid PV-wind systems, and solar PV, have lower operating costs but involve higher capital investment costs relative to diesel systems. However, well-targeted subsidies that aim to reduce the burden of upfront investment cost and/or extend the maturity of supporting financial instruments can facilitate the transition to more sustainable sources of energy. It should be kept in mind that users may not receive the same level of service (available capacity, hours of service per day) from most renewable sources as can be expected from the grid or conventional diesel minigrids.

The provision of renewable energy can be structured as a service delivery program, not merely as an exercise in disseminating a specific technology. Especially for households who live outside grid-accessible areas, renewable energy sources such as solar home systems (SHS) can be an effective option. Their provision can be structured as a service delivery program rather than be focused on a specific technology and simply distributing and installing the systems. The latter approach has yielded uneven results. In contrast, the service delivery program would be akin to a utility service that focuses on regular maintenance and has a measurable means of accountability. A service delivery approach also helps identify priorities in providing electricity access for health care and social and administrative services (table 3). Furthermore, when deployed on a large scale, the service delivery approach can help integrate energy into a wider development strategy for the country.

Incremental investment needs for universal access to electricity are within reach but poorer countries will need external assistance. Figure 6 shows the investment needs for each technology option (grid and off-grid sources) in two electrification scenarios. The incremental investment needs of the Universal Access scenario (with respect to the Business-as-Usual scenario) are US$26 billion, or US$1.3 billion per year until 2030. This amount is

Table 3. Renewable Energy Options for Off-Grid Areas

Energy service	Existing off-grid energy sources	Examples of new and renewable energy sources
Lighting and other small electrical needs for homes, schools, street lighting, telecommunications, hand tools, and vaccine storage	• Candles • Kerosene • Batteries • Central battery recharging by carting batteries to grid • Kerosene refrigerators	• Hydropower (picoscale, microscale, small-scale) • Biogas from HH-scale digester • Small-scale biomass gasifier with gas engine • Village-scale minigrids and solar/wind hybrid systems • Solar home systems • Solar refrigerators
Communications (televisions, radios, mobile phones)	• Dry cell batteries, • Central battery recharging by carting batteries to grid	• Hydropower (picoscale, microscale, small-scale) • Biogas from HH-scale digester • Small-scale biomass gasifier with gas engine • Village-scale minigrids and solar/wind hybrid systems • Solar home systems
Process motive power (small industry/SMEs)	• Diesel engines and generators	• Small electricity grid systems from microhydro, biomass gasifiers, direct combustion of biomass, and large biodigesters
Water pumping (agriculture and drinking water)	• Diesel pumps and generators	• Mechanical wind pumps • Solar PV pumps • Small electricity grid systems from microhydro, biomass gasifiers, direct combustion of biomass, and large biodigesters

Source: REN21 2010.

Figure 6. Investment Needs in Business-as-Usual and Universal Access Scenarios: Electricity, 2011–30

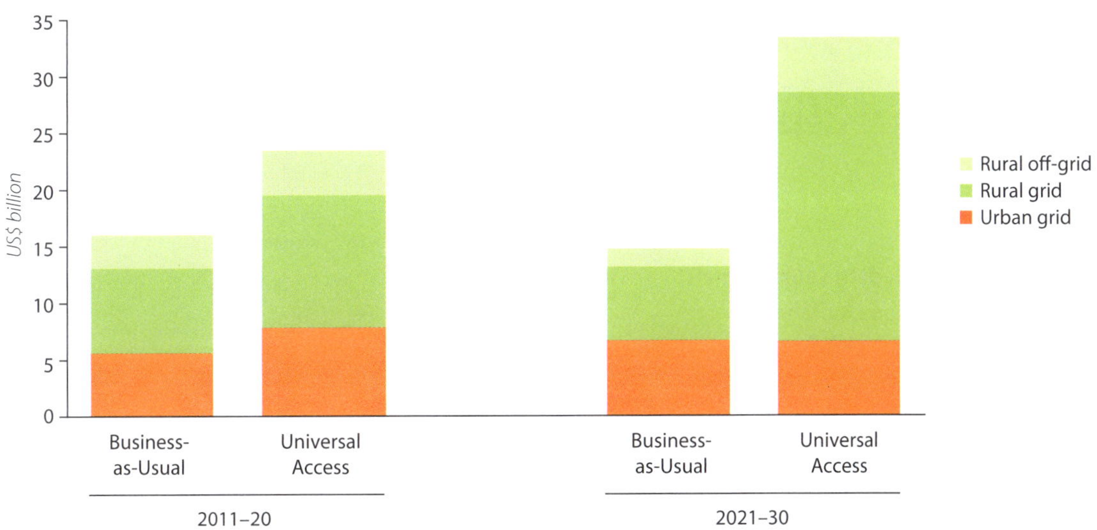

Sources: UN-DESA 2008; authors' calculations.

approximately 0.1 percent of the 2009 Regional GDP (excluding China) and is expected to decline as Regional GDP increases over time. However, in low-access countries (table 1), the incremental investment needs represent a much higher share (1.6 percent) of GDP. This greater share would call for significant concessional financing on the order of US$200 million per year (approximately 1.0 percent of their GDP) to help achieve universal access to electricity by 2030.

Modern Cooking Solutions: Clean Fuels and Advanced Cookstoves

Widespread Use of Solid Cooking Fuels in EAP Countries

The use of solid fuels for cooking is extremely common in EAP, even in countries with high rates of electricity access. Thailand and the Philippines have the highest percentage of people who use modern cooking fuels (mainly LPG but also electricity; and biogas in rural areas) for cooking. However, even in Thailand, well over 33 percent of the people use wood, straw, or charcoal as their main cooking fuel. China has by far the largest absolute number of people who lack access to modern cooking fuels, including approximately 400 million people who use coal,[5] whose quality varies significantly from region to region.[6] Coal also is commonly used for cooking in Lao PDR, Mongolia, and Vietnam. Other countries in the Region—Cambodia, Indonesia, and the Philippines—depend heavily on biomass fuels such as wood or agricultural waste for cooking. Seventy percent to 90 percent of the people in these countries use biomass fuels for cooking.

Increasing Access to Modern Cooking Fuels: Significant but Differing Challenges between Rural and Urban Areas in EAP Countries

There are significant differences between the patterns of urban and rural use of modern cooking fuels, solid fuels (coal and charcoal, wood, straw, and animal dung), and kerosene (figure 7). The countries that face the greatest challenges in increasing the use of modern cooking fuels and advanced cookstoves for solid fuels are Cambodia, Lao PDR, and Mongolia, in both urban and rural areas. Solid fuels also are relatively more dominant in the rural areas of Indonesia, the Philippines, and Vietnam. Kerosene accounts for a large proportion of cooking fuel in Indonesia.

Two Scenarios for Access to Modern Cooking Fuels and Advanced Cookstoves: Business-as-Usual and Universal Access

Similar to the exercise for electricity access, this report has developed two scenarios for cooking fuel use in the urban and rural areas of EAP. These scenarios take into account the projected increases in population and patterns of rural-urban migration up to 2030.

The Business-as-Usual scenario assumes that there will be no growth in the percentage of HH using clean cooking fuels in either urban or rural areas. This conservative assumption is based on the fact that none of the EAP countries has in place concrete policies to accelerate access to modern cooking fuels or advanced cookstoves. However, under the Business-as-Usual scenario, due to increasing urbanization, the absolute number of people using modern cooking fuels would increase. In rural areas, by the year 2030, 665 million people still would be using solid fuels for cooking, 460 million of whom would be dependent on wood straw or dung. In contrast, in urban areas, under the Business-as-Usual scenario, approximately 350 million people would be dependent on solid fuels, including 150 million people who would be using coal and 100 million people who would be using other solid fuels (figure 8).

The Universal Access scenario for cooking fuels assumes that the Region's entire urban populations would use modern cooking fuels by 2030. This assumption is based on the fact that the main barrier to urban universal access to modern cooking fuels is limited affordability—and this barrier could be overcome by an appropriate

5. Government of China 2009.

6. Aden and others 2009.

Figure 7. Patterns of Cooking Fuel Use in EAP Countries, 2009

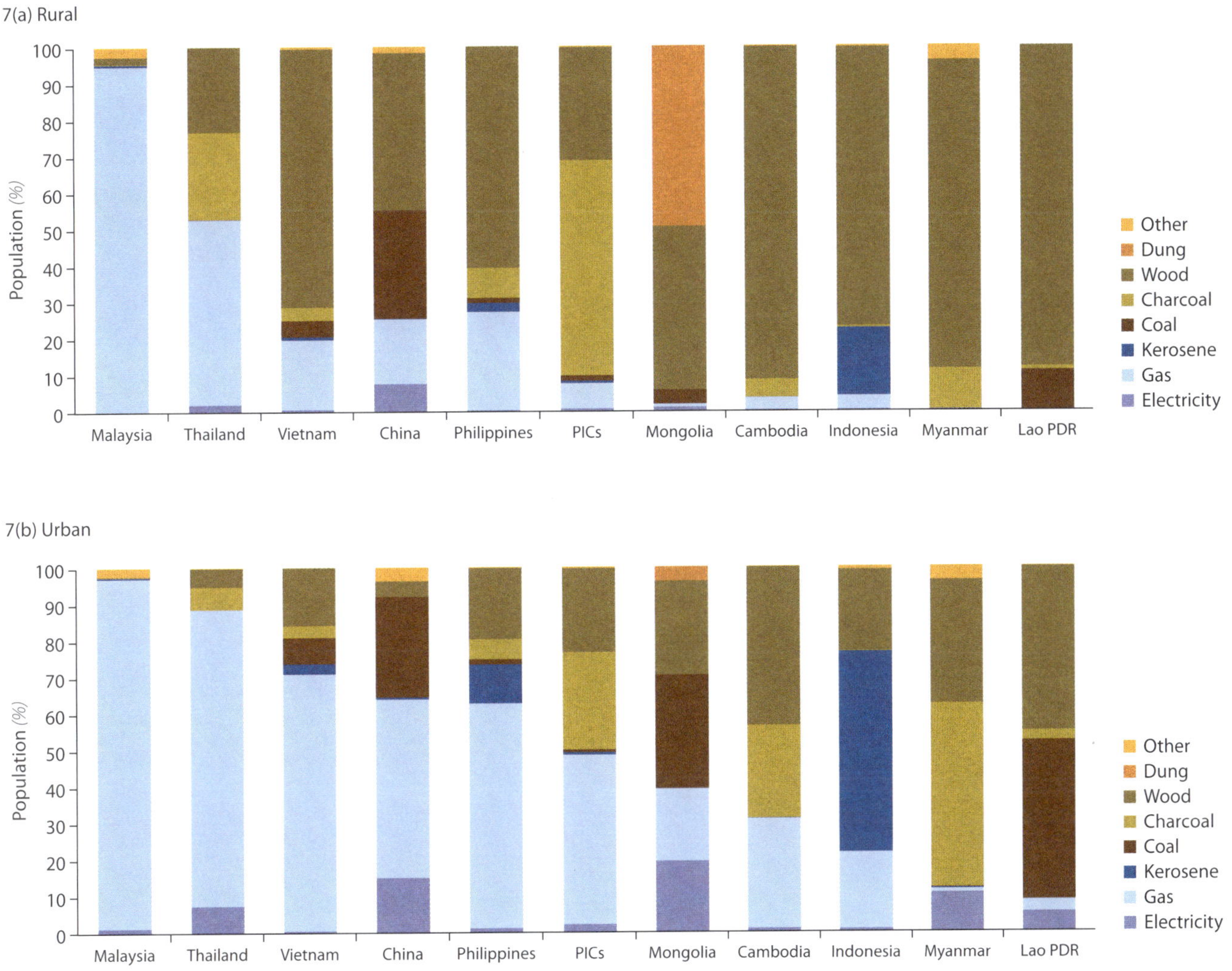

Source: WHO and UNDP 2009.

Note: The percentages are based on HH surveys conducted primarily 2006–08. Gas included biogas, piped gas, and LPG.

use of subsidies. However, the rural population would switch only partially to modern cooking fuels due to the lower affordability and lack of availability of LPG. The Universal Access scenario assumes that, across the entire Region, there would be a 20 percent increase in the use of modern fuel use compared to the Business-as-Usual scenario. Under the Universal Access scenario, by 2030, in rural areas, 300 million people still would be using traditional fuels, and over 130 million people would be cooking with charcoal or coal (figure 8). The promotion of advanced, clean-burning, high-efficiency stoves is an essential component of the Universal Access scenario for modern cooking solutions.

Modern cooking fuels and advanced stoves are key to clean and efficient cooking in EAP countries. In urban areas, achieving universal access to modern cooking fuels will depend primarily on the expansion of natural gas and LPG use. In rural areas, the path to clean and efficient cooking will require higher penetration of biogas and the universal application of efficient stoves for solid fuels.

Figure 8. Use of Cooking Fuels in EAP by Urban and Rural Populations, 2010–30

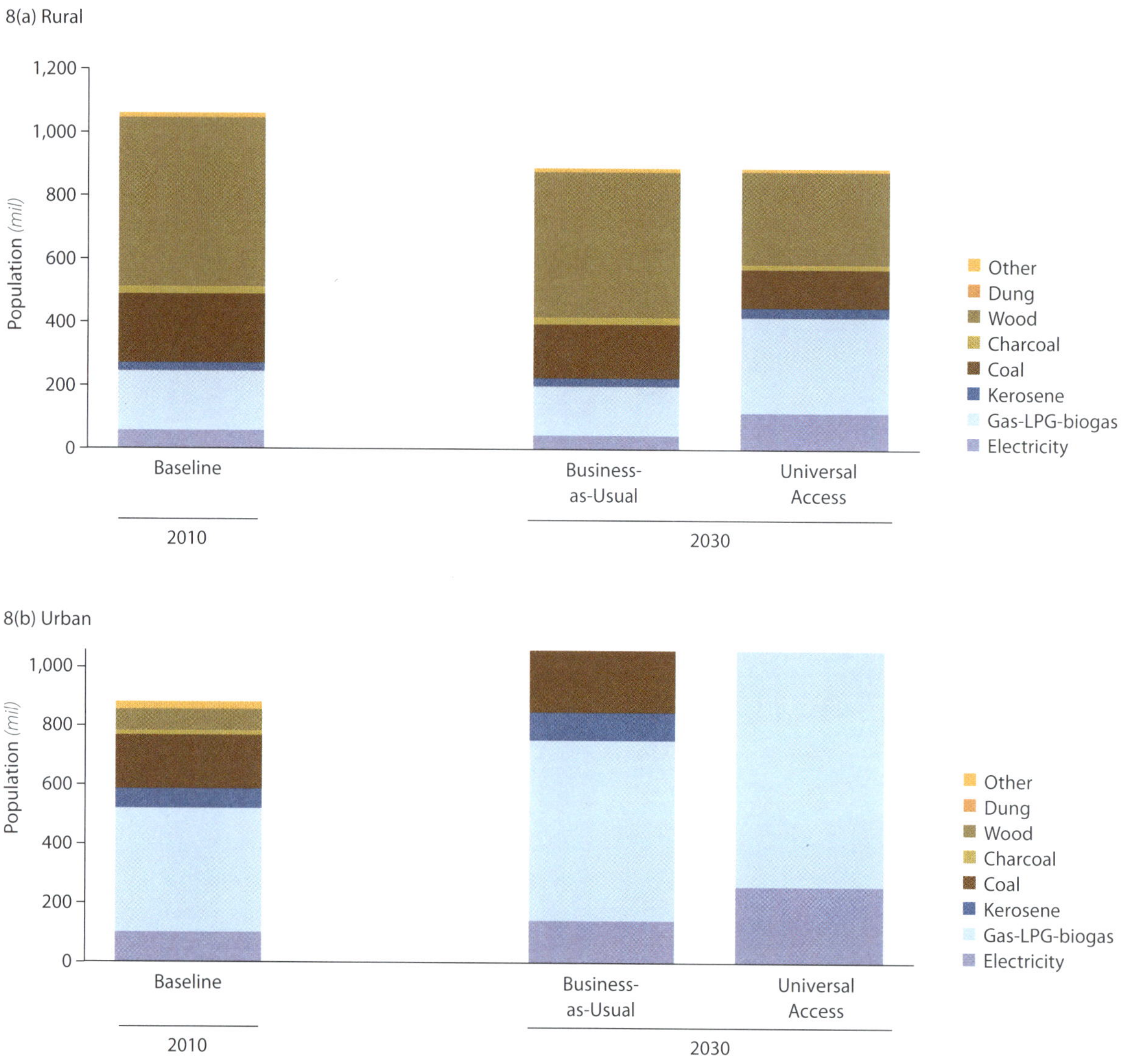

Sources: WHO and UNDP 2009; UN-DESA 2008; authors' calculations.

It will be more difficult to expand the infrastructure for natural gas and LPG to serve customers in remote rural areas, in which biogas may be the main option for a modern cooking fuel for HH with adequate livestock.

In urban areas, the main challenge is to develop the infrastructure for modern cooking fuels. This challenge will require significant investments in importing and processing facilities for natural gas and LPG, in addition to investments in the electricity infrastructure.

In rural areas, the major challenge is the deployment of efficient stoves. The main challenge in rural areas will be to develop and market large numbers of advanced stoves that burn coal and traditional biomass. Mass-scale dissemination of these stoves faces four major barriers:

1. EAP in general exhibits a lack of momentum among national institutions to promote advanced cookstoves.
2. The channels for providing credit for the manufacture or purchase of advanced

stoves, and for marketing them widely, are underdeveloped.
3. There are no widely accepted standards or certifying institutions to qualify the stoves as safe, durable, efficient, and clean burning.
4. Financial support for the technical development of advanced stoves is limited.

Financial incentives are essential to encourage the transition to clean and efficient cooking. The adoption of modern fuels and advanced cookstoves depends on three key factors: income level, pricing policy, and physical access to fuels. As their incomes rise, HH in EAP can be expected to switch to LPG and various specialized electric cooking appliances. However, past income growth in EAP has not been enough to move the majority of the people to modern fuels, even in urban areas, which have physical access to modern fuels. Therefore, it is necessary to introduce pricing policies that will accelerate the transition to modern fuels or improved stoves. Such pricing policies also are needed to attract the investments in physical infrastructure required for market expansion.

Fuel taxes and subsidies have major impacts on energy consumption patterns for cooking. EAP countries employ a wide range of policies to tax or subsidize cooking fuels. These policies often are related to a country's natural resource endowments. If a country is required to import a fuel, it is likely to be taxed. If the fuel is produced within the country, price subsidies are more likely. For example, in Indonesia, kerosene subsidies have been high in the past. Consequently, it is the only EAP country in which kerosene is used extensively in both rural and urban areas for cooking. In China, 30 percent of people in both urban and rural areas use coal for cooking,[7] as it is readily available and has a relatively low price. The problem with heavily subsidized prices is that the cooking fuel inevitably is purchased for use in other sectors such as transport and industry. Thus, subsidies for cooking fuels should be set below the level that makes these fuels commercially attractive for non-cooking uses, and taxes should not be set so high as to make the fuel too expensive to use for cooking.

Access to cooking fuels in some EAP countries is related to import policies. Until the early 1990s, China limited the import of LPG to conserve foreign exchange. This policy meant that LPG was informally rationed, and even people who could afford it could not actually purchase it.[8] In addition, many of the smaller EAP countries such as Cambodia and Lao PDR also limit the import of petroleum-based fuels, which means that the majority of their populations remain dependent on solid fuels for cooking.

Reducing the upfront cost of LPG cookstoves is important to accelerate use of LPG. Poor people often cannot afford the upfront cost of an LPG stove even if they can afford to buy the fuel. One clear policy solution is to subsidize the purchase cost of the stove. Likewise, LPG cylinders that often are sold in containers that hold a month's supply of fuel may be too expensive for poor HH, who are used to buying small amounts of fuel several times a month. The solution is to adopt suitable cooking fuel import policies and perhaps provide subsidies or loans to the poor to pay the upfront costs of both modern and advanced solid fuel cookstoves

A major emphasis on marketing and promoting biogas energy systems is needed in rural areas. For farmers that keep animals, biogas energy systems can be an efficient way to increase access to modern fuels. Such biogas systems can provide clean

7. WHO and UNDP 2009. Numbers provided by industrial and commercial sources may differ from this estimate, which is based on HH surveys.

8. Beginning in the early 1990s, as its LPG market was opened to competition from international investors, China's LPG consumption increased rapidly by an annual growth rate of 18%. By the end of the 1990s, China became the third largest LPG consumer, after Japan and the US. China's LPG consumption is highly dependent (40%) on imports. The country will continue to face challenges in the development of its LPG market (Tian 2002).

gas for cooking and have been successfully promoted in China and Vietnam. However, they still are not widely known or disseminated in other EAP countries.

Most important, there is a need to develop advanced cookstoves that use traditional fuels. Under a cookstoves program in China, over 100 million HH have adopted improved cookstoves. However, even the best of these stoves are based on outdated technology that does not achieve the combustion efficiency associated with modern fuels. For modern cooking solutions, the development and promotion of advanced, clean-burning, high-efficiency stoves are prerequisites for the Universal Access scenario.[9]

Financing Requirements for Clean Cooking under the Business-as-Usual Scenario

Under the Business-as-Usual scenario, the investment requirement for clean cooking would be approximately US$16 billion by 2030 (figure 9). The financing requirement for rural areas would be approximately US$6.6 billion by the same year, including US$4.6 billion for wood stoves and US$1.7 billion for coal stoves. In urban areas, the investment requirement by 2030 would be approximately US$9.5 billion, including US$1 billion for wood stoves, US$2.5 billion for coal stoves, and US$4.4 billion for advanced fuel stoves.

Financing Requirements for Clean Cooking under the Universal Access Scenario

Under the Universal Access scenario for clean cooking, the total financing requirement would be approximately US$22 billion by 2030 (figure 9). The additional investment for urban areas by the same year would be approximately US$4 billion, and for rural areas approximately US$2 billion.

One Goal: Achieving Universal Energy Access

Universal Access under both the electricity and clean cooking paths is affordable, and both must be implemented. In the past, the EAP countries have paid more attention to electricity than to clean cooking, as have large donor and multilateral institutions, including the World Bank. However, the one goal of universal modern energy access can be achieved only by implementing *both* the electricity *and* the clean cooking paths. The combined incremental investment in the Universal Access scenario for electricity and modern cooking solutions will be approximately US$32 billion over 2010–30, a small fraction of Regional GDP. The benefits from achieving universal access are significantly greater than the investment costs.

Developing a Regional entity to facilitate universal energy access to "bottom-of-the-pyramid" HH.[10] At present, most EAP countries share the common problem that they are not well positioned to provide HH energy systems nor improved cookstoves to the approximately 15 million HH (approximately 75 million people) who will be out of the range of the main grid or supply chains of modern fuels. The products suitable for these HH, efficient cookstoves in particular, have not yet been fully developed, tested, and certified. Furthermore, the financing schemes suitable for the low and variable levels of income of these HH have not yet been established. These two challenges are related, and to a large degree shared, by EAP countries. Therefore, after consultation with the EAP countries and donors, it is proposed to establish a Regional forum or entity whose objective would be to facilitate the knowledge sharing and capacity building in improving access to modern energy in poor rural areas. This Regional entity also would serve as a focal point via which to channel the assistance of various interest groups, such as NGOs that are interested in improving the quality of lives of the "bottom-of-the-pyramid" HH. The Regional entity would not substitute for the necessity to improve institutional and regulatory capacity of the countries that already have relevant institutions, or to build institutional capacity where it does not exist. In

9. Smith and Deng 2010; Sinton and others 2004.

10. "Bottom-of-the-pyramid" refers to the poorest socioeconomic group, often defined as those living on less than US$2 a day.

Figure 9. Investment Needs for Modern Cookstoves under Business-as-Usual and Universal Access Scenarios by Cooking Fuel

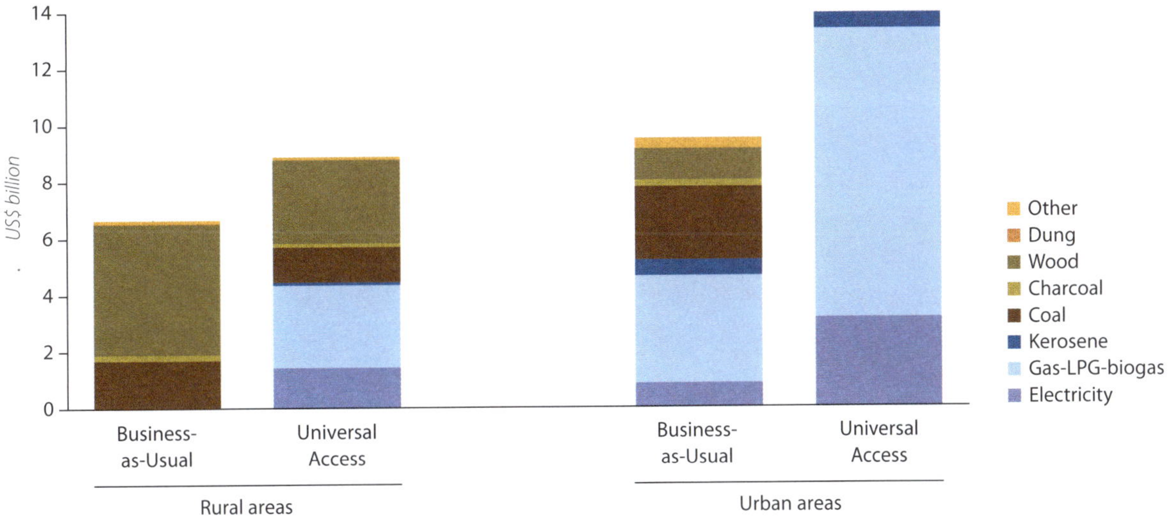

Sources: WHO and UNDP 2009; authors' calculations.

fact, the Regional entity would be effective only through close collaboration with national institutions and promotion of their joint learning and implementation of best practices.

Regardless of the approach, there needs to be a credible international entity that ensures that the issue of universal energy access gets the attention that it deserves at the Regional level.

It would be essential that any new Regional forum have the endorsement of the beneficiary countries. Hence, this report recommends that this issue be discussed with the EAP countries concerned and that additional steps be taken in consultation with them.

Solar panel on a ger, Mongolia.

Photo credit: Ashden Awards, Flickr.

1

Two Paths to Universal Energy Access in the EAP Region

In the past three decades, the countries of the East Asia and the Pacific Region (EAP) have experienced the fastest economic growth in the world. This growth was accompanied by rapid urbanization. As a consequence, EAP energy consumption has more than tripled and is expected to double again over the next two decades (World Bank 2010e; IEA 2010). The 2010 World Bank report, *Winds of Change: East Asia's Sustainable Energy Future*, highlighted issues that relate to sustainable energy and development in the middle-income EAP countries. The other—and perhaps even more pressing—challenge faced by all countries in the Region is to provide access to the hundreds of millions of people who still lack the most basic access to electricity and modern cooking solutions.[11] Even among the EAP countries that have progressed in this regard, there often are wide variations in access between urban and rural areas. The lack of access to modern energy services touches on a number of developmental concerns, ranging from income growth to environmental and health issues. The purpose of the current flagship report is to address energy access and related developmental issues in EAP that so far have received less attention compared to the macro-energy issues of climate change and reduction of greenhouse gas (GHG) emissions.

EAP countries have two steep paths to climb to achieve universal access to modern energy: Electricity and Modern Cooking Solutions. Approximately 170 million people, or 34 million households, in EAP countries do not have electricity connections in their homes. This number is equivalent to approximately 9 percent of the Region's total population, and 30 percent of the Region's population excluding China. Moreover, approximately 6 times that number, or over 1 billion people, still lack access to modern cooking solutions. To put these figures in perspective, among all the Regions served by the World Bank, EAP has the largest number of people who do not have access to modern cooking fuels. In addition, EAP is exceeded by only Sub-Saharan Africa and South Asia in the number of people who lack access to electricity (figures 1.1a and 1.1b).

Energy Access, Poverty, and Development

The relationship between the use of modern energy and economic development is fairly well established. For instance, it is widely accepted that electricity use and GDP per capita are highly correlated (figure 1.2).

Energy and development are mutually reinforcing factors. Access to energy not only results

11. "Modern cooking solutions" refers collectively to modern cooking fuels (electricity, LPG, and biogas) and advanced, clean, and efficient stoves for coal, charcoal, wood-based biomass, and dung.

Figure 1.1 Population without Access to Modern Energy

1.1(a) Population without electricity access, 2009

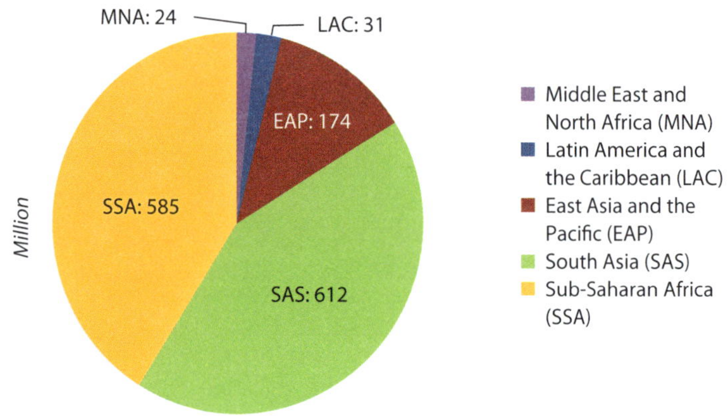

Sources: IEA 2010; authors' estimates.

1.1(b) Population without access to modern cooking fuels, 2007

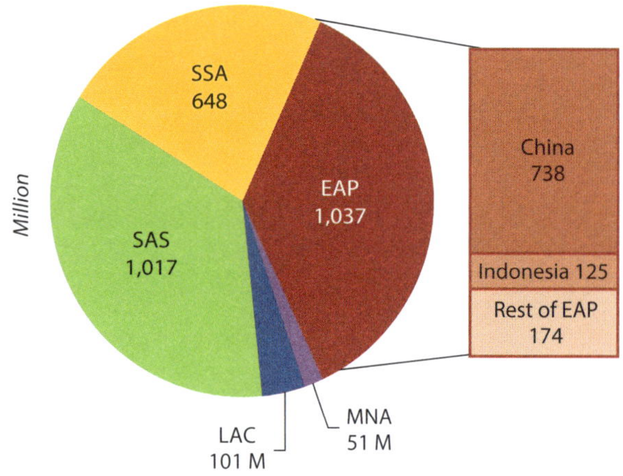

Source: WHO and UNDP 2009.

Figure 1.2 GDP per Capita versus Electricity Use, 2008

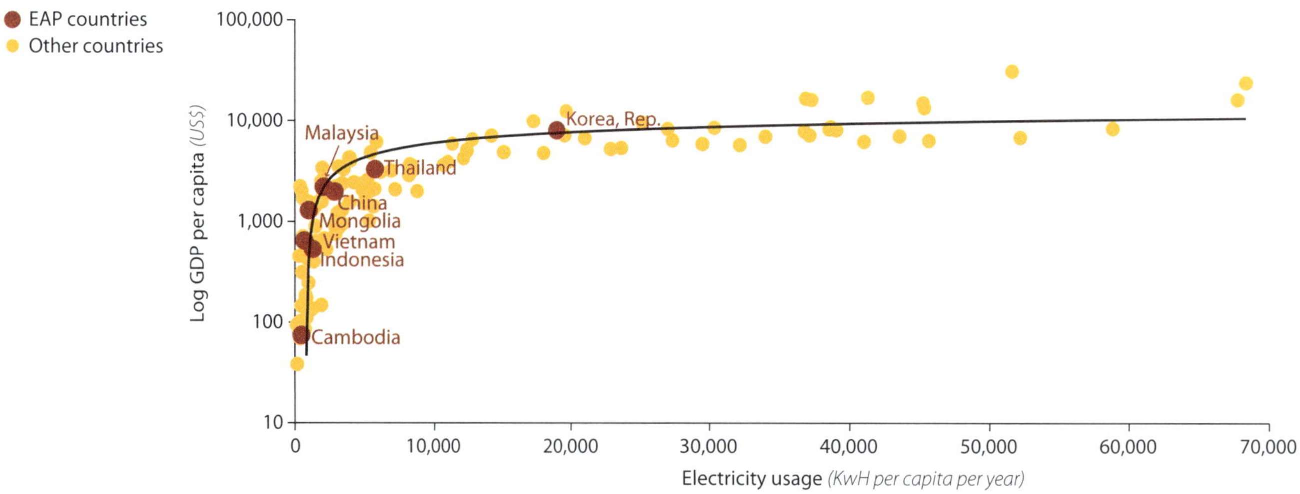

Source: World Bank 2010e.

from, but also actively contributes to, economic growth (IEA 2010). Access to modern forms of energy is essential, first, to provide clean water, sanitation, and health care. Second, access produces sustainable development benefits by supplying reliable and efficient lighting, heating, cooking, mechanical power, and transport and telecommunication services. As incomes rise, access to electricity tends to grow faster than access to modern cooking fuels, largely because governments tend to pay more attention to electrification (IEA 2010). EAP is no exception.

Box 1.1 Importance of Modern Energy in Achieving Millennium Development Goals (MDGs)

Goal 1. Eradicate extreme poverty and hunger. Access to modern energy facilitates economic development by providing more efficient and healthful means to undertake basic HH tasks and the means to undertake productive activities. Modern energy can power water pumping, thus providing drinking water and increasing agricultural yields through the use of machinery and irrigation.

Goal 2. Achieve universal primary education. In impoverished communities, children commonly spend significant time gathering fuelwood, fetching water, and cooking. Access to improved cooking fuels or technologies facilitates school attendance. Electricity also facilitates communication and education, particularly through information technology but also through providing such basic needs as lighting.

Goal 3. Promote gender equality and empower women. Improved access to electricity and modern fuels reduces the physical burden associated with carrying wood. Access also frees valuable time, especially for women, widening their employment opportunities. In addition, street lighting improves the safety of women and girls at night.

Goals 4, 5, and 6. Reduce child mortality; Improve maternal health; and Combat HIV/AIDS, malaria, and other diseases. Most staple foods require cooking. Reducing HH indoor air pollution through more healthful cooking fuels and stoves decreases the risk of respiratory infections, chronic obstructive lung disease, and lung cancer (from burning coal). Electricity and modern energy services support the functioning of health clinics and hospitals.

Goal 7. Ensure environmental sustainability. Modern cooking fuels and more efficient cookstoves can relieve pressures on the environment caused by the unsustainable use of biomass fuels. The promotion of low-carbon renewable energy is congruent with the protection of the local and global environment. Using cleaner energy also reduces greenhouse gas emissions and global warming.

Source: IEA 2010.

However, access to both electricity and modern cooking solutions is essential to address the enduring impacts of poverty and to move the poor onto a rising development trajectory.

Modern Energy and the Millennium Development Goals

The link between access to modern energy and development is most clearly defined by the Millennium Development Goals.[12] The MDGs were formulated to reduce global poverty while increasing education, empowering women, and improving child and maternal health. Although there is no direct reference to energy in the MDGs, the need for access to energy, particularly modern energy, to improve overall welfare is well recognized by the development community. The importance of access to modern energy to achieve each of the MDGs is described in box 1.1.

Of particular importance among the MDGs are the profound implications of improving the health and reducing the workload of women and children. The use of solid fuels (coal, charcoal, wood-based biomass, straw, and dung) for cooking predominates in the Region, even in countries that have significantly enabled access to electricity. Women and children in particular are exposed to indoor cooking smoke from incomplete burning of biomass fuels in inefficient stoves, in the form of particulates up to 20 times higher than the maximum levels considered safe by WHO (WHO 2005). Strong evidence of

12. Agreed during the 2000 United Nations Millennium Summit, the Millennium Development Goals (MDGs) set forth the most important development goals for the world community in poverty reduction, health, education, and environment. The MDGs were set to be achieved by 2015.

causal linkages between indoor biomass combustion and family health (such as acute respiratory illnesses, heart disease, cataracts, and even cancer) is now available (Kammen and others 2002; Parikh and others 2001; Smith and others 2004). Smoke from solid cooking fuels is estimated to result in approximately 665,000 premature deaths annually in EAP (of approximately 2 million such deaths in developing countries worldwide) (WHO and UNDP 2009). In the absence of a suitable policy framework that sustains market-based solutions, these numbers are expected to rise in keeping with trends in developing countries worldwide. This trend contrasts with other leading causes of premature deaths (HIV/AIDS and malaria), which are expected to decline given the continued support of active programs (IEA 2010) (figure 1.3).

Access to electricity also has a significant impact on the achievement of MDGs. Lighting is one of electricity's most important and widely adopted benefits and positively impacts many of the MDGs. Lighting enables reading after daylight and a better environment for education, development of home enterprises, social interaction, and leisure activities. Lighting a single 60-watt bulb for 4 hours a day can produce as much as 100 times more light than traditional kerosene lamps, and even more when compared to candles. The improved lighting and use of electricity to refrigerate clean water also improves health and creates opportunities for more productive activities over a longer period during evening hours. A recent study in rural Bangladesh indicates that access to electricity has the cumulative impact of increasing rural HH incomes by as much as 20 percent, resulting in a corresponding drop in the poverty rate of approximately 15 percent (Barnes and others 2010).

Figure 1.3 Exposure to Indoor Air Pollution from Biomass in Developing Countries: A Major Health Issue, 2008–30

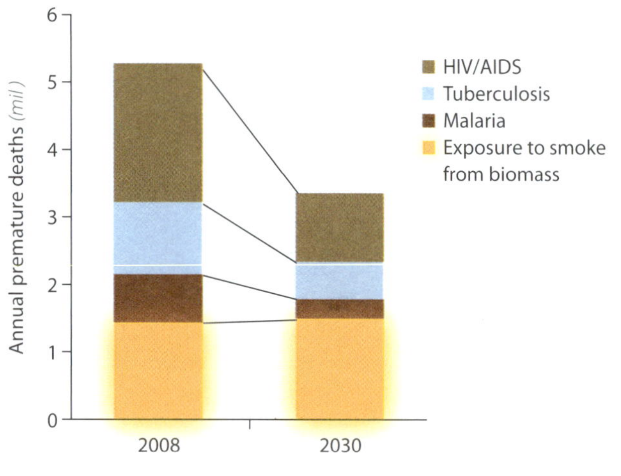

Source: IEA 2010.

Beyond Access: Affordability and Reliability of Electricity Service

If the expected social and economic benefits from electricity are to be realized, electricity service needs to be affordable and reliable.[13] Electricity usage is likely to claim a higher share of rural incomes than of urban incomes. When electricity connection and usage costs are perceived to be unaffordable, HH may choose not to connect despite being in an area that already has access to electricity. Furthermore, compared to urban areas, rural electricity supply is likely to be subject to more frequent and longer service interruptions and lower overall quality of supply. If faced with low affordability and poor reliability of service, HH will display greater reluctance to pay even for existing services. This would put pressure on the revenues of the electricity provider, leading to further deterioration in service. Seen in this context, not merely providing electricity access but also ensuring the affordability, reliability, and quality of electricity supply is crucial for sustainable benefits.

Minimal Impact of Universal Access to Modern Energy on Climate Change

Meeting the goal of universal access to modern energy will have only a marginal impact on climate change. Reaching the goal of universal access to modern energy services is expected to have a relatively small impact on energy demand

13. The report focuses on providing access at the HH level through grid or off-grid means as appropriate and feasible. It does not make any assumptions relating to the level of consumption per HH.

Figure 1.4 Poverty and Lack of Access to Electricity by Country, 2008

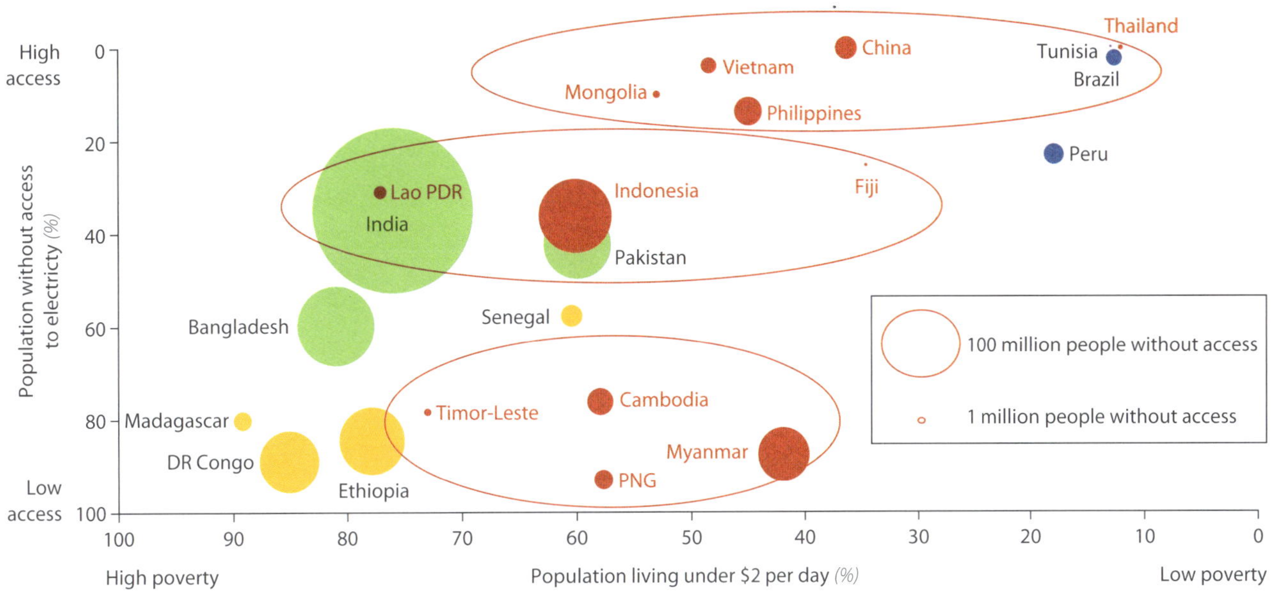

Sources: IEA 2010; World Bank 2010e; authors' estimates.

and production, and CO_2 emissions.[14] Switching to modern cooking fuels and using efficient cookstoves also would mitigate climate change by reducing the CO_2 emissions and deforestation associated with the use of traditional fuels. According to a recent estimate (IEA 2010), achieving universal electricity access would have a modest impact on energy-related CO_2 emissions, raising them by 0.6 percent–0.8 percent (approximately 2 percent of current OECD emissions) by 2030. Another estimate suggests that, under the most unfavorable assumptions, providing basic electricity access to the world's unconnected HH would add only 0.33 percent to global GHG emissions, and much less if renewable energy and efficient light bulbs could be deployed (IEG 2008). The welfare benefits of electricity access are on the order of US$0.50–US$1 per kilowatt hour (kWh). In contrast, a stringent valuation of the corresponding carbon damages in a worst-case scenario is far less at only a few cents per kWh (Gilbert 2009).

Electricity Access in EAP: Success, Stagnation, and a Rural-Urban Divide

The last decade has seen major progress in increasing electricity access in several EAP countries, but the overall picture is still a mix of success stories and unmet challenges. In the Region and the across the world, electricity access rates tend to be inversely correlated with poverty rates (figure 1.4). However, some EAP countries have been remarkably successful in increasing electricity access to higher levels than their poverty rates would predict. In this respect, in recent years, Lao PDR and Vietnam stand out both in the Region and on the global stage. On the other hand, Cambodia, Myanmar, and PNG are at lower levels of electricity access than might be expected based on their poverty rates.

14. A related issue concerns aerosol emissions from biomass cookstoves, which consist of both black carbon and organic carbon. While the emissions characteristics of biomass burning in cookstoves are considered critical for climate science, surprisingly little concrete scientific data exists on such key factors as the ratio of organic carbon to black carbon. This ratio is critical for calculating the effect of HH biomass combustion on global climate models. As a result, there is still significant uncertainty about whether black carbon emissions from burning biomass in cookstoves have a net warming effect on climate globally.

Table 1.1 Electricity Access in EAP Region, 2009

Country	Electricity access 2009 (%)	Population without electricity 2009 (mil)
Indonesia	65	81.4
Myanmar	13	43.9
Philippines	84	15.0
China	99	8.0
Cambodia	24	11.4
PNG	7	6.3
Vietnam	96	3.6
Lao PDR	70	1.9
Timor-Leste	22	0.9
Thailand	99	0.7
Mongolia	90	0.3

Sources: IEA 2010; authors' estimates.

Overcoming the Income Barrier to Electricity Access

Despite the broad correlation globally between relatively high poverty and low access to electricity, the experience of Lao PDR and Vietnam within the past decade shows that it is possible to break out of this pattern and potentially leverage electricity access for economic development. China and Thailand were the first to reach near-universal access to electricity (95 percent or greater) over a decade ago. Among the medium-access countries (access rates greater than 50 percent), Indonesia and the Philippines have made reasonable progress over the last decade in the face of population growth. Nevertheless, in these 2 archipelagic countries, 96 million people of 317 million (approximately 30 percent) still lack access to electricity, primarily in rural and remote island areas. Figure 1.5 shows that overcoming the income barrier to electricity access goes hand-in-hand with accelerated economic development in EAP. This synergy is particularly important for low-access countries such as Cambodia and Papua New Guinea (PNG).

Rural-Urban Divide in Electricity Access in EAP

Many EAP countries show a vast rural-urban divide in access to electricity. However, others have shown the way to overcome this divide. Among the middle income EAP countries, Indonesia has the largest gap between urban and rural access (urban

Figure 1.5 Growth in Electricity Access versus GDP per Capita in EAP, 2000–09

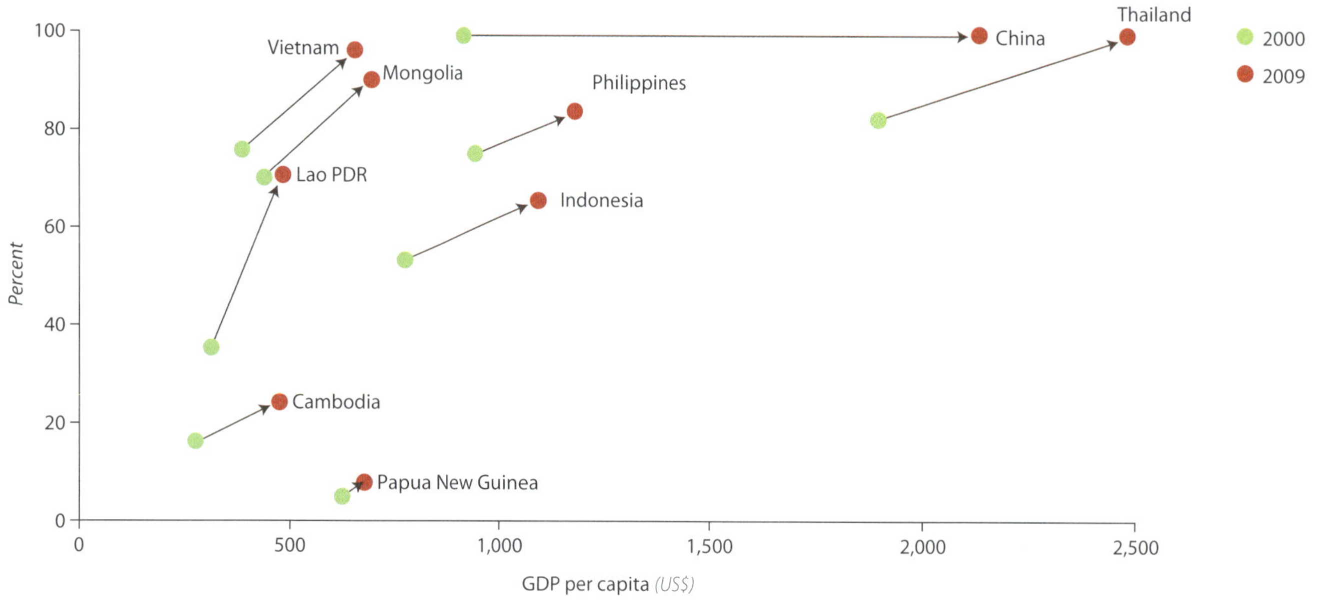

Sources: IEA 2010; World Bank 2010e; authors' calculations.

access: 97 percent; rural access: 32 percent). This gap arises due partly to the geographic remoteness of its people, who live in the country's many distributed island *kabupatens,* or districts. Among the lower-income EAP countries, Lao PDR has made considerable progress in reducing the gap between urban and rural access but faces increasingly bigger challenges as the grid extends to distant and hilly terrain.

Nevertheless, the experience of China and Thailand and, more recently, Vietnam shows that the challenges of connecting areas with low population density and difficult terrain can be overcome through government commitment backed by appropriate institutional and financial policies. Each of these countries has achieved viable electrification of most of its rural areas (figure 1.6). For the archipelagic countries, recent developments in renewable technologies, including HH energy systems, increasingly are providing viable technical alternatives to grid-based electrification. Over the past two decades, such approaches are being successfully used in China and Thailand to cover the "last mile" of universal access to electricity.

Figure 1.6 Rural-Urban Divide in Electricity Access in EAP Countries, 2008

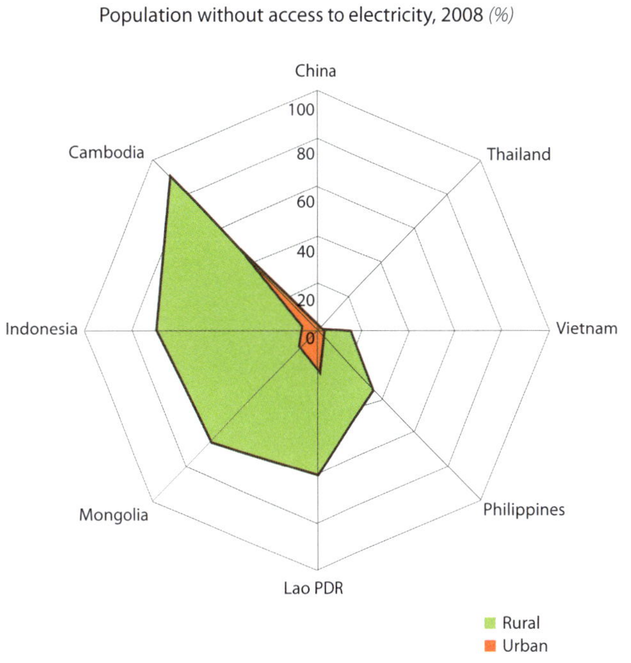

Sources: IEA 2009; authors' calculations.

Widespread Use of Solid Fuels in EAP

In the EAP Region, well over 50 percent of the population relies on solid fuels for cooking. This majority contrasts markedly with the EAP record in electricity access. Although countries such as China, Thailand, and Vietnam have set up extensive programs to address electricity access issues, they continue to rely heavily on solid fuels for cooking (figure 1.7). Thailand has the Region's highest rate of access to modern cooking fuels (mainly LPG). Nevertheless, well over 33 percent of its population uses wood, straw, or charcoal as the main cooking fuels.

In all EAP countries, whether in rural or urban areas, it is usually the poor HH who are likely to lack access to modern cooking fuels. This dichotomy is particularly visible in urban areas. There, a large majority of HH have access to electricity; nevertheless, large numbers still must resort to solid fuels for cooking (and, in some countries, for heating) since they are more affordable relative to LPG and electricity. Thus, many urban residents continue to use coal, wood, straw, and dung despite their negative impacts on human health and the environment (through both indoor and outdoor air pollution).

In rural areas, as could be expected, the proportion of people who cook with solid fuels is higher than in urban areas (table 1.2, figure 1.8). However, the health implications of cooking with these fuels using traditional stoves are similar for both rural and urban areas. In contrast to urban areas, most of the solid fuels burned in rural areas, mainly wood-based biomass, are not purchased but are collected from the local ecosystem.

Laying Out the Energy Access Challenges for EAP Countries

Varying Challenges at Different Levels of Electricity Access

With respect to electricity, EAP countries can be classified into countries with high, medium, and low access. China, Thailand, and Vietnam have achieved high or near-universal access (95 percent of HH or more), primarily through grid

Figure 1.7 Poverty and Lack of Access to Modern Cooking Fuels: EAP and Other Countries, 2007

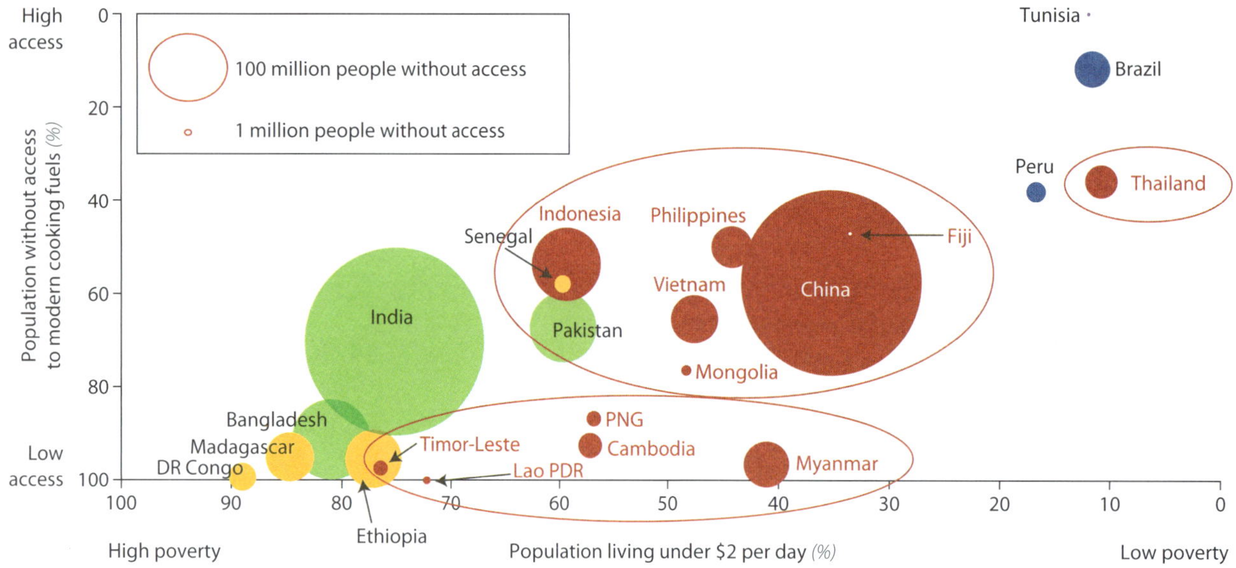

Sources: WHO and UNDP 2009; World Bank 2010e.

Table 1.2 Population without Modern Cooking Fuels in EAP, 2009

Country	Population without modern cooking fuels (%)	(mil)
China	58	768
Indonesia	54	124
Vietnam	66	57
Myanmar	97	47
Philippines	51	45
Thailand	37	25
Cambodia	93	15
PNG	87	6
Lao PDR	97	5
Mongolia	77	2
Timor-Leste	100	1

Sources: WHO and UNDP 2009; authors' calculations; Government of China 2008; Carolina Population Center and NINFS 2008.

Figure 1.8 Rural-Urban Divide in Use of Modern Cooking Fuels in EAP Countries, 2007

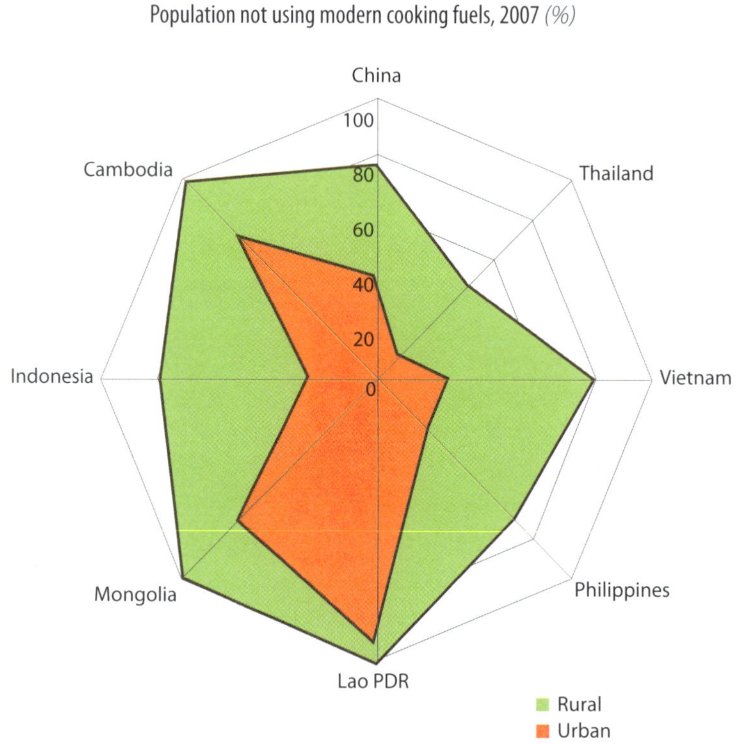

Sources: WHO and UNDP 2009; authors' calculations.

extension. The remaining electrification efforts in these countries are focused on the more remote and sparsely populated areas. Indonesia, Lao PDR, Mongolia, and the Philippines are in the medium access range (50 percent–95 percent of HH). In these countries, the challenge is to maintain and/or accelerate their momentum in expanding electricity access in the face of increasing costs. Finally, the countries with low access (below 50 percent)—Cambodia, Myanmar, and most of the PICs—have yet to significantly initiate their national programs to increase access to electricity and consolidate their institutional framework and sector regulation. Therefore, the nature of challenges facing the EAP Region in increasing electricity access varies significantly among countries with low, medium, and high access.

The high electricity-access countries—China, Thailand, and Vietnam—have benefited from strong government commitment, planning, and local involvement. Of China's 1.3 billion people, approximately only 8.0 million remain without electricity (table 1.1). With its population of approximately 65 million people, Thailand now provides electricity to virtually all HH who can possibly be connected to the electricity grid. Its remaining population is on track to be provided with off-grid—primarily renewable—electricity in the near future. The factors that have enabled these countries to provide electricity to their rural and more remote populations are a sustained commitment by their governments over a long period, as in the case of China (box 1.3); the dedication of their service providers in carefully planning and implementing programs; and engaging local governments to facilitate the electrification programs in their respective areas.

China's and Thailand's successes are closely followed by those of Vietnam, which has made major progress in extending grid-based electricity access to its rural areas. With a population of close to 96 million people, Vietnam has

Box 1.2 China: Steady but Slow Transition to Modern Cooking Fuels

Box figure 1.2.1 Households cooking mainly with LPG or electricity in China, 1989–2006

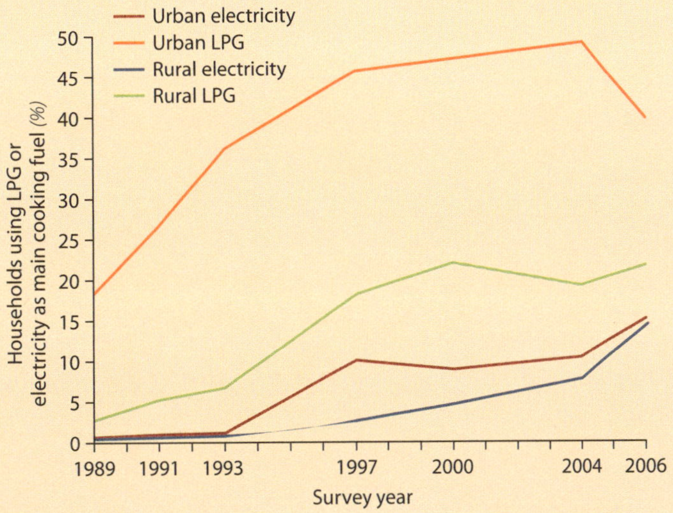

Despite China's impressive accomplishments in economic growth and poverty reduction, Chinese households have been slow to climb the "energy ladder" and transition from using coal and traditional fuels to modern cooking fuels. Even among high-income groups, HH continue to use coal and biomass for cooking, and/or a diverse portfolio of fuels to meet their different energy needs. The transition to modern fuels is faster when access to reliable electricity has been achieved for all sections of society, as illustrated by the slow but steady growth of LPG use for cooking in China's urban areas (box figure). However, despite progress in transitioning to modern fuels over 15 years—during which period China's GDP per capita (current US$) has increased nearly 700 percent—the majority of people in China still cook with solid fuels.

Source: WHO and UNDP 2009; Government of China 2008; Carolina Population Center and NINFS 2008.

Box 1.3 China: Sustained Commitment to Universal Electrification

China's experience with rural electrification is unique among developing countries in that it grew from decentralized efforts, rather than as a centrally driven program. From relatively small and scattered beginnings in 1949 aided by progressive support from the government, and more recently driven by a market-oriented approach, China has managed to bring electricity to over 99 percent of its rural population. The following factors contributed to China's successful electrification program.

- ***The government*** has played a major role in rural electrification through laying down overall policy for rural electrification, providing financial support, promoting diverse financing sources, and developing technical standards and support.
- ***Power sector institutions*** have evolved from central planning to more private investment, competition, and less regulation.
- ***Financial and operational viability*** is ensured through subsidized sources of finance from central and provincial governments that increasingly are supplemented by private sector lending and capital.
- ***Reliability and safety*** were addressed through a major reform of the rural power management system, which began in 1988. At one stage, the reform improved reliability to 99.7 percent, and the losses in the low voltage (LV) grids fell by 30 percent–45 percent.
- ***Costs*** were kept under control by developing two compatible national technical codes—one for established regions and the other for poor and remote regions, with provision for a smooth transition.
- ***Off-grid electrification efforts,*** beginning in 1996, have brought electricity to 1.3 million people through photovoltaic (PV) and PV/wind hybrid systems.
- ***The poor have been served*** through favorable connection-cost policies, and direct assistance has been provided to the most vulnerable.
- ***Large-scale renewable energy*** is being ramped up with recently announced plans to support the deployment of 500–600 MW of large-scale solar PV in both on-grid and off-grid areas by 2012.

Source: Barnes 2007.

approximately only 2 million people who lack access to electricity. After having connected approximately 60 percent of its population by the late 1990s, Vietnam successfully reorganized its electricity sector.[15] This was done in part to address the growing problems faced by rural electricity distribution companies as they faced increasing difficulties in maintaining the pace of electrification and service quality. As a result, Vietnam has reached a high access rate (96 percent in 2009) while improving service performance and reducing costs.

In medium-access countries, the challenge is to maintain the momentum of the electrification effort. Several EAP countries have reached the majority of their populations but still have significant challenges ahead of them if they are to reach the goal of universal access to electricity. The largest challenges facing the two archipelagic countries—Indonesia and the Philippines—arise from the relatively high cost of extending electricity access to the large rural population that lives on remote islands and in hilly areas. Lao PDR faces increasing costs as it continues to expand electricity access to increasingly remote areas. However, its electrification program shows remarkable vitality, as demonstrated in the last 15 years of rapid electrification. With the lowest population density in the world and a significant nomadic population scattered over 1.5 million square miles, Mongolia faces unique challenges in

15. Appendix 1: Vietnam.

Table 1.3 Electricity Access Challenges in EAP

Level of electricity access *(% HH)*	Grid	Off-grid
High access *(>95)* China, Thailand, Vietnam	Finalizing "last-mile" issues	Innovating energy solutions for remote HH
Medium access *(50–95)* Indonesia, Lao PDR, Mongolia, Philippines	Maintaining momentum of programs; jumpstarting programs that have stagnated	Solidifying existing efforts and making necessary reforms to serve communities and HH in remote areas
Low access *(<50)* Cambodia, Myanmar, most Pacific Island countries (PICs), and Timor-Leste	Getting started and making a serious commitment to expand national grid	Developing the institutional and regulatory framework for off-grid solutions

Source: Authors.

providing electricity access to its dispersed population. These challenges will need to be addressed through innovative solutions.

Weak institutions, lack of planning, and low ability to subsidize rural electrification are common in low-access countries. Although EAP countries with low electricity access rates—Cambodia and most PICs—have diverse characteristics, they share three traits: weak institutional capacity, lack of integrated planning for a national electrification program, and low ability to finance capital-intensive rural electrification programs from fiscal resources. Among this group, the Pacific Island Countries (PICs) are further challenged by remoteness, large dista nces separating locations, and low population densities. In many of the PICs, the extension of grid electricity is virtually impossible. They must depend on innovative solutions and new technologies typically associated with off-grid electrification programs.

The electricity access challenges for EAP countries are summarized in table 1.3, which also maps the main topics and issues covered in this report regarding electricity.

Modern Cooking Solutions: Modern Cooking Fuels and Clean and Efficient Stoves

In urban areas, lack of availability of modern fuels is the main factor limiting access to clean and efficient cooking. Access to modern cooking fuels (mainly LPG) often is related both to import and pricing policies and to the affordability of the stove and fuel. In the 1990s, China's policies limited the imports of LPG. These policies resulted in an informal "rationing" of LPG so that even people who could afford to switch to LPG could not actually purchase it. This situation has improved in recent years. Several other EAP countries, notably Cambodia, Lao PDR, and PICs, also restrict the imports of modern cooking fuels to different extents, mainly to preserve foreign exchange for other strategic imports.

Rural areas need to put a major emphasis on marketing and promoting new efficient solid fuel stoves and biogas systems. During the 1980s and 1990s, China successfully implemented the National Improved Stove Program (NISP) in which over 100 million people adopted improved cookstoves. An additional constraint is the lack of availability of efficient technologies for cookstoves that use biomass. Such advanced cookstoves are particularly needed in rural areas, which are unlikely to gain access to modern cooking fuels in the near future. While manufacturers have begun producing newer high-efficiency stoves, cost reductions through large-scale production still have not been achieved to make these affordable to rural consumers, who have limited purchasing power. These consumers continue to depend on obsolete stoves that are polluting and inefficient. For rural areas,

Table 1.4 Context and Challenges in Promoting Modern Cooking Solutions in EAP

	Income and affordability	Fuel prices and subsidies	Access
Urban	• Higher incomes and greater affordability mean greater use of modern fuels by high-income groups. • Transitional or traditional fuels are used by lower income groups.	• Fuel prices are significantly affected by tax and subsidy policies. • Traditional fuel prices often follow the prices of modern fuels so it is important not to excessively tax modern fuels, which indirectly would raise cooking costs for the poor.	• Availability of modern fuels and transition fuels is low. • Fuel import limits mean formal or informal rationing in certain countries. • Transition and biomass fuels still are used by many of the poorest people.
Rural	• Lower incomes and lower affordability mean greater use of transitional and traditional fuels. • Excessive collection time for biomass fuels and health problems (from indoor pollution) further reduce productivity and HH income.	• Modern and transitional fuels are priced beyond affordable levels for low-income consumers. • Implicit price of biomass fuels often is high if based on collection time. • Where available, market price of biomass is relatively high compared to modern fuels based on energy content.	• Modern and transitional fuels often are not distributed in rural areas. • Traditional fuels often are overharvested, reducing their availability and increasing collection distances. • Biogas systems programs are small in most countries and relevant only for farmers with livestock. • Advanced stoves are still in development and pilot stage and require refinement.

Source: Authors.

China's and Vietnam's experiences demonstrate the potential of expanding biogas energy systems to HH who own livestock.

The above issues and key challenges for increasing access to clean and efficient cooking in EAP countries are summarized in table 1.4.

Following this chapter's overview of energy access in EAP, the remainder of the report focuses on key issues and strategies for increasing access in the Region.

Chapter 2. ***Designing a National Electrification Program for Universal Access*** discusses the key principles that guide the development of national electrification programs and how they apply in the EAP countries. This chapter also compares two scenarios for electricity access: (1) the Business-as-Usual scenario, based on the current plans and electrification trends in EAP countries; and (2) the Universal Access scenario, under which all HH will have access to electricity by 2030.

Chapter 3. ***Electricity Access: Delivering Results on the Ground*** explores the policies that are essential to successfully implement national electrification programs, particularly in the context of rural electrification.

Chapter 4. ***Modern Cooking Solutions: Status and Challenges*** lays out the main issues in promoting modern cooking solutions in the two scenarios—Business-as-Usual and Universal Access—and compares their implications for urban and rural consumers.

Chapter 5. ***Modern Cooking Solutions: The Way Forward*** looks at the policies and strategies that are essential to increase access to clean and efficient cooking fuels and methods, for both rural and urban areas.

Chapter 6. ***One Goal: Achieving Universal Energy Access in the EAP Region*** looks at the synergy between two access paths—electricity and modern cooking solutions—in achieving the common goal of universal access to modern energy solutions in EAP countries by 2030.

Appendix 1. ***Electricity Access: Selected Country Briefs*** summarizes key issues for increasing electricity access in low- and medium-access EAP countries.

Appendix 2. ***Energy Access Projects Funded in the EAP Region by IBRD, IDA, and GEF, 2001–10*** lists the energy and electricity projects that have been in the World Bank's and Global Environment Facility's (GEF) portfolios for Energy Access in EAP since 2001.

Replacing ad hoc electricity generation systems (top) with efficient distribution networks (bottom).

Top: Run of the river pico-hydro generators, Lao PDR.
Bottom: Distribution networks in rural areas, Vietnam.

Photo credits: Top—Openyourap, Flickr; Bottom—Hung Tien Van.

2

Designing a National Electrification Program for Universal Access

Achieving universal access to electricity presents unique challenges to each EAP country in its own social, political, and cultural context. This report has identified two categories of EAP countries that have yet to provide electricity access to a significant portion of their populations. These are the *low-access countries*—Cambodia, Myanmar, and most PICs; and the *medium-access countries*—Indonesia, Lao PDR, Mongolia, and the Philippines. As these countries develop their strategies and implementation plans to expand electricity access, they have the advantage of being able to draw on the practices of countries in EAP and around the world that already have achieved near-universal electricity coverage. Successful national electrification programs share three key principles that have governed their planning:

1. ***Cost-effectiveness.*** Technical solutions that are used to provide a specified level of reliability and quality of electricity service should be selected on a least-cost basis.
2. ***Affordability.*** Electricity service should be affordable to consumers, fiscally affordable to the subsidy provider, and affordable to the service provider vis-a-vis its financial viability.
3. ***Timeliness of implementation.*** Electricity access should be provided to the intended beneficiary in a timely manner.

This chapter examines each of these three principles for planning electricity access. The chapter unbundles the issues that are of most relevance to EAP countries in the context of the overall vision of achieving universal electricity access in EAP by 2030.

Cost-Effectiveness: Least-Cost Technical Options for Desired Reliability and Quality of Service

Grid-based electricity has been the mainstay of electricity access. The experience of most EAP countries and other parts of the world shows that grid-based electricity is the mainstay of electricity access. The extension of the electricity distribution grid generally is the least-cost approach to reach new consumers and to rapidly increase the number of HH with access to electricity. In most EAP countries, more than 90 percent of electrification has been achieved through grid extension both in urban and rural areas (figure 2.1). Some EAP countries stand out with a relatively lower share of grid-based electricity due to their demographic and geographic characteristics. For instance, in Mongolia, almost 33 percent of the population consists of nomadic herder families; and off-grid electricity is their main option. The other exceptions, due to their remote and dispersed locations and their sparse populations, are the PICs and the outlying archipelagic regions of Indonesia and the Philippines.

Figure 2.1 Estimated Shares of Grid-Based and Off-Grid Electricity in EAP Countries, 2009

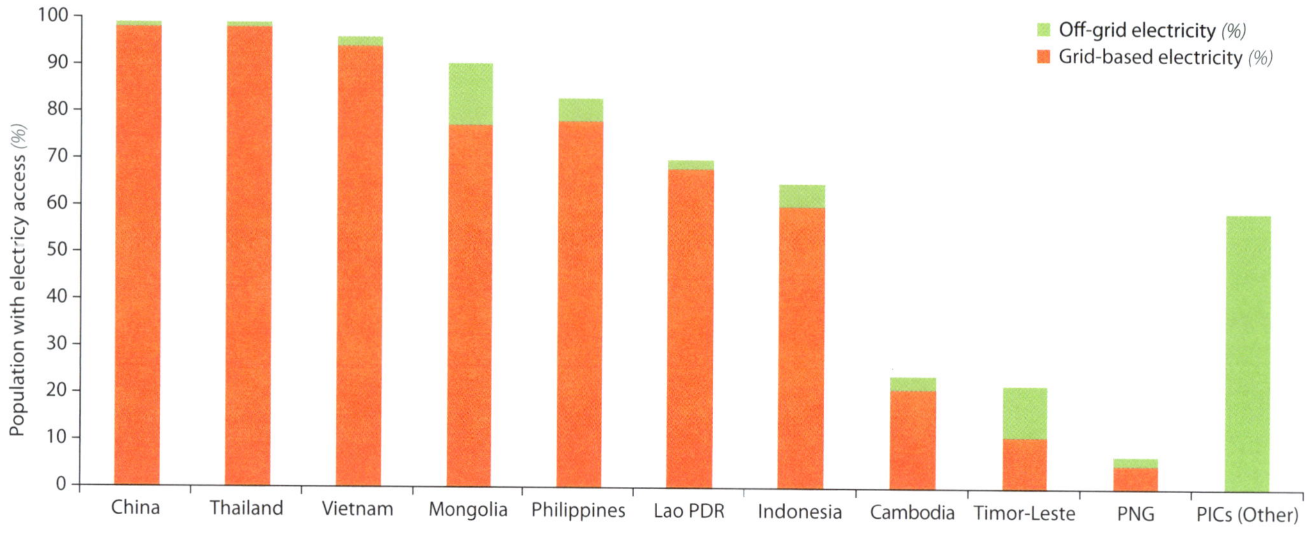

Sources: IEA 2010; authors' estimates.

Achieving universal access in a least-cost manner requires use of both grid and off-grid technologies. Until approximately two decades ago, extension of the main grid, diesel-powered minigrids, and minihydropower generators were, in most circumstances, the only electrification options available to rural communities. However, the emergence of new, and primarily renewable, energy technologies has opened up new possibilities for expanding electricity service to areas in which the grid-based electricity is not viable or could take a long time to reach. With the commercial maturation of various small-scale, renewable-energy-based technologies, new off-grid sources—from solar photovoltaic (PV) systems to small wind generators and microhydropower—have become viable alternatives for increasing electricity access. These sources are especially appropriate in remote and dispersed communities or for HH whose consumption levels are very low and are expected to grow slowly. However, in recent years, successful countries have adopted strategies that include both grid and off-grid approaches (figure 2.1). Such a multitrack approach is based on several factors including the cost of supply, projected electricity demand, and the expected development impacts of electrification.

Providing electricity through the main grid becomes progressively more expensive as the grid spreads outward from the urban and periurban into rural areas. The higher cost is due to decreasing economies of scale due to lower population densities in the outlying areas and the longer distances over which the grid must be extended. In the face of increasing construction costs per consumer and the logistical difficulties and associated costs encountered in managing rural systems, providing electricity by extending the grid gradually becomes less cost effective.

The costs of grid extension can become prohibitively high in mountainous areas and regions that are difficult to access, such as remote and dispersed islands, all of which make up a significant part of several EAP countries. In 2003 in Lao PDR, electricity access levels were approximately 50 percent, and the average cost per HH connection through grid extension was US$450–$550.[16] Today Lao PDR has a 70 percent access level, and the average cost per HH has doubled to approximately US$900 as the grid enters more remote rural areas (World Bank 2010c). This cost is bound to increase as the grid pushes into hilly terrain.

16. Connection cost includes cost of extending the MV and LV lines and providing the service drop.

In rural Peru, it has been estimated that the unit cost of providing connectivity through the main grid could be up to five times higher than in cities (World Bank 2010b), due to the higher capital costs of reaching remote and mountainous areas.

There is a wide range of off-grid technologies that can serve individual HH or small communities. Sources of off-grid electrification can be classified in two broad categories: (1) isolated minigrids, which can serve communities with up to several hundred or even thousands of HH; and (2) HH energy systems, which typically serve one HH. In between are community systems that generally are scaled-up HH systems that serve small schools, health centers, or community facilities. The technologies used for HH systems can range from small gasoline/diesel generators to solar, wind, and picohydro systems, all of which have a typical rating of 50W. Community systems can have a higher capacity of 300W or more. Minigrids can be powered by a wider range of technologies including diesel (by far the most common), minihydro, municipal solid waste/landfill gas, geothermal, biomass gasifier, solar-thermal with storage, and wind-based systems.

Grid and off-grid energy options have a wide range of costs. Table 2.1 presents the "levelized power generating costs" for various technologies used for HH energy systems and isolated minigrids.[17] The table compares them with the costs for the large-scale conventional sources that typically power the main grid. Table 2.1 shows that the range of levelized costs for HH systems (US$0.15–$0.65) is distinctly higher than that for minigrid systems (US$0.7–$0.51).[18] In contrast, the large generation sources that typically power the main grid provide the cheapest electricity (US$0.4–0.23).

Generally speaking, grid systems are least cost when they serve large loads in areas of high population density. Isolated off-grid systems are least cost for serving smaller, localized loads in places

Table 2.1 Levelized Power-Generating Costs for Various Technologies, 2005

Range of applications	Generation source	Rated output *(kW)*	Levelized cost *(USc/kWh)*
Household/ community systems	Pico/microhydro	0.30	15
	Wind	0.30	35
	PV-wind hybrid	0.30	42
	Solar PV	0.05	62
	Diesel/gasoline generator	0.30	65
Isolated minigrids	Biogas	60	7
	Minihydro	5,000	7
	Municipal: Solid waste/ landfill gas	5,000	7
	Geothermal binary	20,000	7
	Biomass gasifier	20,000	7
	Biomass gasifier	100	9
	Diesel base load	5,000	9
	Pico/microhydro	100	11
	Solar-thermal with storage	30,000	13
	Geothermal binary	200	16
	Diesel/gasoline generator	100	20
	Wind	100	20
	PV-wind hybrid	100	31
	Microturbines	150	32
	Solar PV	25	51
Main grid	Coal steam subcritical	300,000	4
	Large hydro	100,000	5
	Combined cycle natural gas	300,000	6
	Oil steam	300,000	7
	Oil	300,000	12
	Combustion turbines: Natural gas	150,000	13
	Combustion turbines: Oil	150,000	23

Source: World Bank 2007b.

17. Levelized costs are the present value of the total cost of building and operating a generating plant over its financial life converted to equal annual payments and amortized over expected annual generation from an assumed duty cycle.

18. The figures shown in table 2.1 should be considered indicative, since new technological developments are likely to have changed the pattern of costs, particularly for several off-grid renewable systems, which, in recent years, generally have shown a downward trend.

far from the grid. Similarly, HH systems such as solar PV systems are least cost for low-load areas that are even farther from the grid and in less accessible areas. Figure 2.2a shows schematically the cost-effectiveness boundary between isolated grids and solar PV systems as the boundary changes with HH density and the number of HH served. Similarly, figure 2.2b shows how the cost-effectiveness boundary between grid electricity service and off-grid solar photovoltaic (PV) changes for the same parameters (HH density and number of HH served) as the distance to the grid varies. The figure uses 2 cases: the population served being 3 km and 5 km from the existing main grid. The implication of these figures is that the cost-effectiveness of the technical option chosen for electricity expansion is driven by the density and clustering of the consumers. The lower the values are in combination, the greater the case for individual solar PV. In addition, the greater the distance of the beneficiary population from the grid, the more attractive solar PV becomes.

Figure 2.2 Schematic Profiles of Relative Costs of Electricity Supply: Main Grid versus HH Solar PV Systems

2.2(a) Cost curve for HH solar service in villages remote from main grid

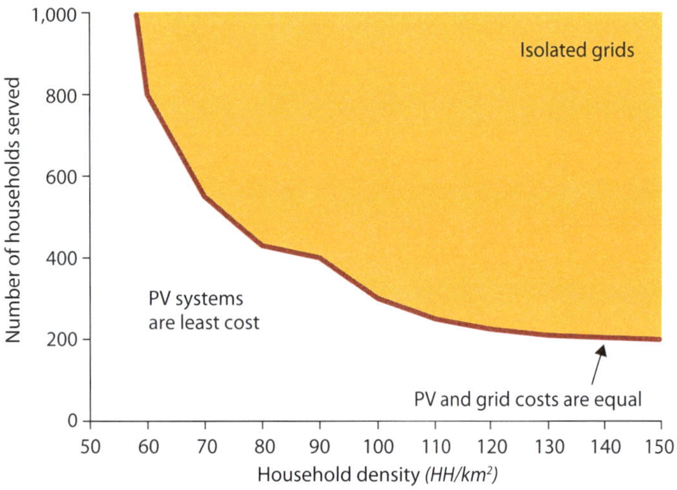

2.2(b) Cost curve for HH solar service in villages 3–5km from main grid

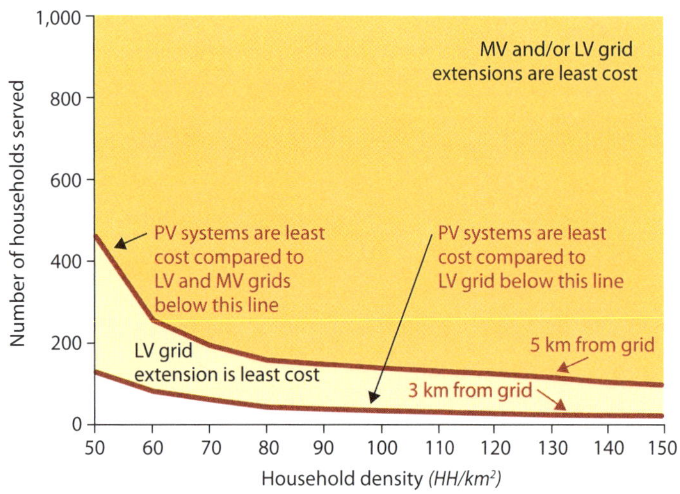

Sources: World Bank 2008c; authors' calculations.

Low-cost solutions and uniform technical standards can help stretch the reach of the grid within the same resource envelope. Incorporating low-cost solutions and uniform technical standards in the planning and design stages can substantially reduce construction and procurement costs. Such advance planning can significantly expand the grid's reach and coverage by stretching the available resources. Thailand's Provincial Electricity Agency (PEA) followed a comprehensive strategy of reducing costs that included (1) standardization of technical features and all equipment and components used for construction of distribution systems; (2) centralization of procurement process and bulk purchases; and (3) reliance on locally manufactured materials, which often were cheaper than imported materials (Barnes 2007).

Lao PDR uses low-cost concrete poles that are manufactured locally and deploys low-cost pre-stressed concrete pole technology. Institutionalizing such innovative cost-saving measures significantly lowers the cost of both fabrication and transportation to the construction site. The simpler technology and production process makes it feasible to produce the poles in portable "mobile factories" that are easily dismantled and relocated to follow the progress of construction as the network spreads out farther. These innovations are a significant achievement because pole costs typically represent approximately 33 percent of the total investment costs of rural electrification programs (World Bank forthcoming 2011).

The savings from integrating low-cost technologies in the early stages of planning and design can reduce construction costs by 20 percent–30 percent without compromising service quality, and can contribute significantly to the pace and scope of electrification programs. If electricity use is likely to be limited to lights and small appliances—a pattern common in rural areas—there may be little justification for applying the same standards as for high-consumption urban areas. Many countries have been successful in reducing construction costs by using technical standards adapted to rural/low demand patterns, frequently adopting low-cost single-phase distribution systems (typically single-wire earth return, or SWER) and centralized procurement processes, and/or incorporating incentives for cost reduction in open and transparent bidding (World Bank 2010a).

For instance, Lao PDR has employed SWER (box 2.1) and has deployed shield wire technology on some HV transmission lines in its mountainous northern region. There, in addition to a more difficult terrain, load clusters tend to be

Box 2.1 Low-Cost Technical Options in Grid-Based Electrification

Single-Phase versus Three-Phase Power Supply

The use of two-wire, single-phase power supply provides several ways of reducing the cost of grid extension for rural electrification. A smaller length of conductor and fewer pole-top assemblies are required, and fewer poles are required between conductors before being limited by clearances. Single-phase supply has been widely used in Brazil, Peru, and Uruguay with up to 40 percent reduction in costs compared to 3-phase configurations. Single-phase lines present no difficulties for HH uses (such as refrigerators or color televisions) or small motors (such as electric pumps or manual tools). However, adaptations and conversions must be made to serve large-motor (above 7.5 horsepower), agro-industrial, and deep-borehole irrigation loads. In practice, most industrial development occurs in industrial zones and incorporated villages, which are supplied with 3-phase, 30-kV lines. Outside these areas, conversion from single-phase to three-phase lines can be made later, if justified by the load.

Single-Wire Earth Return (SWER)

SWER, or single-wire ground return, is a single-wire transmission line to supply single-phase electrical power to remote areas at low cost. It is used principally for rural electrification. SWER is considered to be equally safe, more reliable, less costly, but with slightly lower efficiency, than conventional lines. The SWER line is a single conductor that may stretch for tens or even hundreds of kilometers, visiting a number of termination points. At each termination point, such as a customer's premises, current flows in the line through the primary coil of a step-down transformer to earth through an earth-stake. From the earth-stake, the current eventually finds its way back to the main step-down transformer at the head of the line, completing the circuit.

Box figure 2.1.1 SWER line, New Zealand

Photograph: http://www.flickr.com/photos/16164232@N00/157119903/in/photostream/

SWER's main advantage is its low cost. It often is used in sparsely populated areas in which the cost of building an isolated distribution line cannot be justified. Capital costs are roughly 50 percent of an equivalent 2-wire single-phase line. Several developing nations have adopted SWER systems including Brazil, South Africa, and Tunisia. In Tunisia, use of SWER resulted in a cost savings of 26 percent–30 percent, compared even to single-phase systems. In Lao PDR, EDL has implemented 6 SWER projects in as many provinces, using approximately 135 km of SWER reticulation that connects 4,500 HH as well as a remote army camp.

Sources: World Bank forthcoming 2011; World Bank 2006b; and Barnes 2007.

highly dispersed and small, making it expensive to use normal network design and standards. Shield wire technology enables low-cost tap-offs from the shield wire that runs separately with the HV line, without compromising the security and stability of the high voltage network.

Overextension of the main grid can lower the reliability and quality of electricity supply while increasing losses and cost of supply. A lack of coordinated planning and minimum technical requirements could lead to overextension of the main grid, particularly during periods of rapid build-up. Such systems initially deliver some benefits of electrification. However, they could soon become liabilities due to the decline of service quality, increase in distribution losses, and loss of payment discipline, because dissatisfied consumers would not be willing to pay for poor service.

For example, in the early stages of rural electrification in China, significant safety, reliability, and quality issues surfaced due to lack of coordinated planning and adequate technical standards. In 1988 the government launched a major reform of the rural power management system, and transformed and renovated the rural grids to unify electricity prices for urban and rural dwellers within the same grid (Barnes 2007). Similar problems occurred in Vietnam in the early 1990s, after which the government moved toward better regulation (1998–2004) through defining strategies for the planning, implementation and management of rural electrification (box 2.2). However, the legacy of such an initial phase is a high-cost rehabilitation program that is needed to improve service performance and lower costs. While rolling back grid coverage is very rare, in principle, it may occur due to a combination of factors involving low service performance, high cost, and demographic changes due to migrations from rural to urban areas.

Innovations and new technologies are expanding the scope of off-grid electrification. Technological improvements and declining costs of minigrid and HH energy systems are making off-grid solutions more attractive for rural electrification. Several generating technologies used for off-grid electrification can compete with the cost of electricity provided by the main grid generation options (table 2.1). For example, innovations and mass production of solar PV and small wind generation systems are bringing the cost of these renewable options closer to the cost of conventional off-grid technologies such as microhydro and biogas, which produce electricity more cheaply than do diesel generators. Strong support for renewable energy (RE) under the Kyoto Protocol and related international and national efforts to reduce emissions of GHGs from the energy sector play an important role in increasing investments in R&D and expanding RE technologies. Such investment and expansion can improve their efficiency and drive down costs.

Expansion of telecommunications services provides opportunities for synergy between the electricity and telecommunication business models. The business model used to expand telecommunications services in poor and rural areas often is based on the principle of "slim profit margins and large sales volume," also known as a "bottom-of-the-pyramid," or BoP, model. The BoP model also is potentially attractive in expanding off-grid energy systems. Another synergy between the telecommunications and electrification businesses stems from the need to provide electricity to charge mobile phones as well as to operate cell towers in rural areas. Cellular telecom companies often arrive in remote rural areas faster than do power utilities. The former invest in standalone generation sources, which also can provide surplus electricity to nearby villages. Providing electricity at reasonable prices to rural consumers could help cellular operators increase their "average revenue per user" because rural subscribers are likely to increase as cheap charging facilities become available.

Off-grid household systems have limitations relative to grid-based electricity. There are significant differences in the range and quality of services that can be provided by grid and off-grid sources of electricity. Household energy systems such as solar

Box 2.2 Vietnam: Balancing Speed and Quality in Electricity Grid Expansion

In its push for rapid electrification in the 1990s, Vietnam followed an approach of allowing multiple entities and financing arrangements to construct, manage, and operate rural distribution networks without imposing minimal technical requirements. This approach enabled a rapid increase in electrification: from 1994 to 1997, access increased from 14 percent to 61 percent. However, longer term technical problems regarding reliability and quality of service and distribution losses began to emerge. Furthermore, the low efficiency with which some of these networks were operated, coupled with the lack of financial strength in some of the community rural distribution utilities, undermined the quality and financial sustainability of electricity service.

The government's response to this situation made a profound impact on moving to quality and financially sustainable rural electricity distribution. A significant feature of the response was to define strategies for the system planning, implementation, and management of rural electrification. Furthermore, to enhance the development of the power sector and all electricity activities, the government set a new legal and regulatory framework for the sector. An important recent milestone in this respect was the Prime Minister's Decision 21, issued in 2009, which stipulated a unified national tariff for all residential consumers. It was designed as an incremental block tariff with the first subsidized block to be a lifeline block.[1] The Decision also enabled the takeover by the larger Power Corporations (PCs) of the local distribution utilities that could not demonstrate financial strength. These takeovers consolidated the rural electricity distribution and retail business. In 2010 the Vietnam Distribution Code was approved. It established the rights and obligations of PCs with respect to distribution and retail activities and their customers, including provisions regarding quality-of-service obligations and consumer protection.

As of 2010, it is estimated that 99 percent of the communes and 96 percent of the households in Vietnam were connected to the grid. The four tasks lying ahead are to (1) rehabilitate the low voltage electricity distribution networks in approximately 3,000 communes, (2) determine the most suitable way to achieve the target of electrifying all of the country's households, (3) continue to improve quality of access and reliable supply, and (4) continue to ensure that electricity is affordable to the poor.

Source: World Bank 2011b.

Note:

1. In an incremental block tariff system, a consumer is charged incrementally higher unit rates at higher blocks (range of consumption). "Lifeline block" refers to an allotment of electricity sufficient for a household to cover basic needs, such as lighting, cooking, and water-heating.

home systems (SHSs) provide limited amounts of electricity and service a limited range of applications such as lighting, small appliances (radio, television), and cell phone charging. The use of energy for motive purposes (mechanical power) typically requires more electric power than can be obtained from HH-sized systems. Technically, HH energy systems can be scaled up to almost any size. However, generally, the expense of using several PV cells or installing larger biogas plants or several microhydro systems is too expensive to be considered in a rural context.

The source of electricity has implications for consumption patterns and uses of electricity. For example, in Sri Lanka, the median total wattage of all light bulbs used in a HH was estimated to be 360W for grid-connected HH but only 60W for SHS users. Similarly, HH with SHS own practically no electrical appliances other than a television set, whereas a large proportion of grid-connected HH own a range of appliances (IEG 2008). Initially, SHS and similar HH energy systems can provide a most valuable, albeit limited, amount of electricity in a timely manner.

Box 2.3 Technology Advances in Off-Grid Electrification

Advanced Battery Storage Technologies

Battery storage is one of the key solutions to improve the stability of the power supply from intermittent RE systems such as solar PV. In recent years, battery storage technologies have improved dramatically. Large-capacity systems have been developed that can stabilize power supply from solar or wind systems and provide reliable and quality supplies of electricity for longer periods. For instance, a sodium-sulfur (NaS) battery of 1 MW capacity has been deployed on France's Reunion Island to provide firm power to the island grid system during periods of high demand. The battery is a part of a grid-connected solar PV power station and has an expected lifetime of 15 years. Such storage batteries hold potential for island grids in the PICs and geographically similar regions of Indonesia and the Philippines although, in the near future, investment and operating costs will be a constraint (Drouineau and others 2010).

Getting More Power from Solar PV: Micro-Inverters for Solar Panels

A new type of microinverter is being tested that can dramatically improve the efficiency of solar PV systems. In conventional PV systems, solar panels are wired together in series, and their combined high voltage DC power is fed to a single DC/AC inverter. If one module in the series is affected by a simple cause such as a shadow, a leaf blowing over a module, or dust, the entire array suffers a drop in power output. In contrast, the use of microinverters turns each PV panel into a standalone AC power source and can increase the total amount of power output from a multipanel installation. The use of microinverters is expected to improve a PV system's efficiency by up to 25 percent, while costing approximately 15 percent less than the conventional system. Microinverters can not only help maximize the total power output for a set of solar panels but also make the PV system modular and flexible, enabling expansion. The modular flexibility is an important new feature because the expansion of conventional PV systems requires replacement of the single large inverter, which is one of the most expensive parts of the system. (www.sandfsolar/micro_inverters.html)

DC Microgrids

Due to the fast-increasing number of devices and appliances that require DC power, a new paradigm of the DC household microgrid is emerging. The main idea is to eliminate costs and losses associated with the DC/AC conversion by connecting DC appliances (cell phones, computers) directly to the source of DC power. Typically, a DC microgrid is based on solar PV systems and advanced battery storage, but it also can be connected to the AC network through a converter. At present, the main barriers to adopting this technology for rural electrification are the availability and upfront cost of the microgrid and DC appliances. (Savage 2010)

Nevertheless, in many cases, these systems represent a transitional solution toward a minigrid or integration with the main grid, which can provide a broader range of electrification benefits, particularly in areas that are experiencing fast-growing electricity demand.

Affordability of Electricity Access

The overall viability of a national electricity access program hinges on the affordability of the electricity service to the three main stakeholder groups: (1) consumers, for connection and consumption costs; (2) electricity providers, for their operational and financial viability; and (3) national and local governments, for the fiscal affordability of subsidies needed for sustainable increases in electricity access. These three aspects of affordability are interrelated. All three have been generally secured by EAP countries that already have reached near-universal access. These countries have done so through an appropriate

Figure 2.3 Indicative Electricity Tariffs for Small Customers in EAP Countries

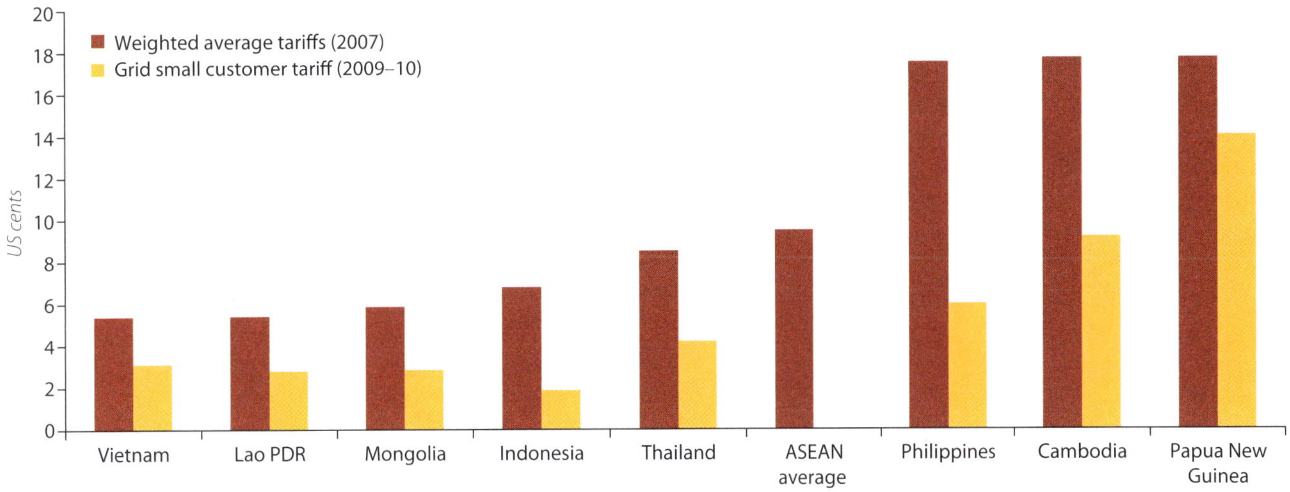

Sources: World Bank 2010c; MIME 2009; National Utility websites: www.evn.com.vn, www.edl-laos.com, www.pln.co.id, www.pea.co.th, www.nea.gov.ph, www.edc.com.kh, and www.pngpower.com.pg

subsidy and tariff policy that makes electricity affordable to consumers; permits cost recovery for the electricity provider; and, where needed, supplements internal generation of funds with external assistance. The bottom line is that the overall cost of increasing electricity access is too high to be fully passed through to newly connected consumers. Consequently, a cost-sharing mechanism is needed. How countries distribute this cost between rate payers and tax payers, and, in some cases, donors providing concessional financing, reflects the, socioeconomic, political, and cultural realities in each country.

Affordable Consumer Tariffs

Most countries seek to make electricity affordable to all consumers through cross-subsidies between two different categories of consumers (industrial and residential) and among different income groups. Targeting the poor and focusing on the customer's access needs are crucial for equity and to ensure that consumers receive an affordable and reliable service that can improve their economic well-being and quality of life. Figure 2.3 shows that the grid-based electricity tariff for the small customer (typically for consumption of less than 50 kWh per month) is significantly lower than the average tariff within most EAP countries.

However, the electricity that is provided by off-grid sources in several EAP countries often is too expensive for the poor customer to afford. For instance, in Cambodia, rural electric cooperative (REC) tariffs range between US20c–$1/kWh, compared to 18c/kWh in Phnom Penh (MIME 2009). In Mongolia, *soums* (district centers) maintain small diesel-powered grids that recover the full cost of expensive diesel fuel. In the PICs, average tariffs are approximately 75c/kWh (appendix 1: PICs). In addition, in Cambodia, Mongolia, and the PICs, supply either is provided for only a few hours per day, or is subject to frequent interruptions. In contrast, Brazil, which has several thousand diesel minigrids in outlying areas, has overcome the issue of affordability through an effective tariff-subsidy system funded from a common source.[19] This tariff subsidy system also has helped secure the financial and operating viability of the electricity providers, which

19. Brazil uses its Energy Development Account (Conta de Desenvolvimento Energético) for making grants to its "Lights for All" rural electrification program. The grant funds come from payments for the use of public assets by private operators, fines collected by the regulatory agency from concessionaires and permit-holders, and annual quotas paid by electricity sellers.

has improved the reliability of the power supply (Niez 2010).

Generally speaking, subsidies are necessary to ensure that the price of electricity is affordable or perceived to be fair in rural areas. However, assessing "willingness to pay" for electricity, particularly in unelectrified rural areas, is not easy. As a result, when the lifeline or some other form of subsidized electricity tariff is designed for poor consumers, underpricing electricity service is a significant risk. The maximum affordable tariff can be determined through HH surveys based on an appropriate assumption that the electricity bill should not be higher than a certain percentage of HH income (for example, 20 percent).

Connection fees present a major barrier to poor HH connecting to electricity. Even in areas that have been provided with access to electricity, many poor HH do not have an electricity connection because they cannot afford the connection fees. The paradox is that even though the individual connection costs and fees represent a very small percentage of the total investment required to expand electricity service, the inability of the consumer to pay for the connection fees keeps the access rate down. When the objective is universal access, poor consumers will require assistance in paying the connection fees.

Connection fees are treated quite differently in many parts of the world (box 2.4). Some HH connecting to grid electricity systems must bear the burden of *grid extension costs* (all investments required to bring the network up to the household). Others must pay only for the *household connection costs* (drop line and the meter that connects their home to the grid). In most cases, customers must pay the full cost at the time of requesting a new connection. Other approaches to ease the burden of the connection fees range from a delayed monthly payment over a relatively long period to a partial or complete subsidy. In Vietnam, average grid extension costs ranged from approximately US$300 in the south—in which communes are more densely populated and MV lines had previously been constructed—to approximately US$400 in northern and central regions.[20] To facilitate affordability, costs to connect communes to the grid have been shared among the Electricity of Vietnam Group (EVN), the provincial governments, the communes and the newly connected customers/households. In the Vietnam Rural Electricity Project I, connection fees required to be paid by HH were less than US$10 per HH, a fraction of the real cost. For the poorest HH, who could not afford even a minimal charge, local authorities often worked out ways to provide assistance. Once connected, however, all consumers were responsible for paying their electricity bills in full.

Ensuring the Financial and Operational Viability of the Electricity Provider

The financial and operational viability of electricity providers depends on full cost recovery of an efficiently run operation while signaling the right incentives to provide good quality services to customers. As the grid expands to outlying areas, electricity providers are faced with increasingly higher costs and a lower revenue base. Under these circumstances, a carefully designed system of subsidies and tariffs that is capable of being administered efficiently is essential to ensure that service in rural areas is not neglected and remains an important focus of the electricity provider. On a larger canvas, promoting productive uses of electricity through small/micro business and services can increase consumption and provide a more stable revenue base for the electricity provider, while increasing the economic impact on the electrified area.

The type and level of subsidies to be applied in a particular country will depend on its social, economic, and political conditions. It often is argued that subsidies for recurring costs should not be encouraged, since these costs grow steadily as access is extended and are not time bound. However, when they are well targeted and applied based on specific monitorable performance standards and criteria, subsidies for distribution/operating costs can be economically efficient and

20. These figures relate to the World Bank's Rural Electrification I and II projects in Vietnam.

Box 2.4 Electricity Connection Fees in Selected Developing Countries Including EAP

Electric connection fees are treated quite differently in many parts of the world. Some HH connecting to grid electricity systems must bear the burden of infrastructure extension costs. Others must pay only for the drop line and the meter that connects their home to the grid. In most cases, customers must pay the full fees at the time of requesting a new connection. However, some electricity companies provide loans, to be repaid over several years with the monthly bill to new customers. The goal of all companies should be to keep initial costs for new customers as low as possible so that they will attract new customers who can take advantage of the benefits of electricity and become sources of revenue to the electricity distribution company.

Box figure 2.4.1 Indicative electricity connection fees in selected developing countries

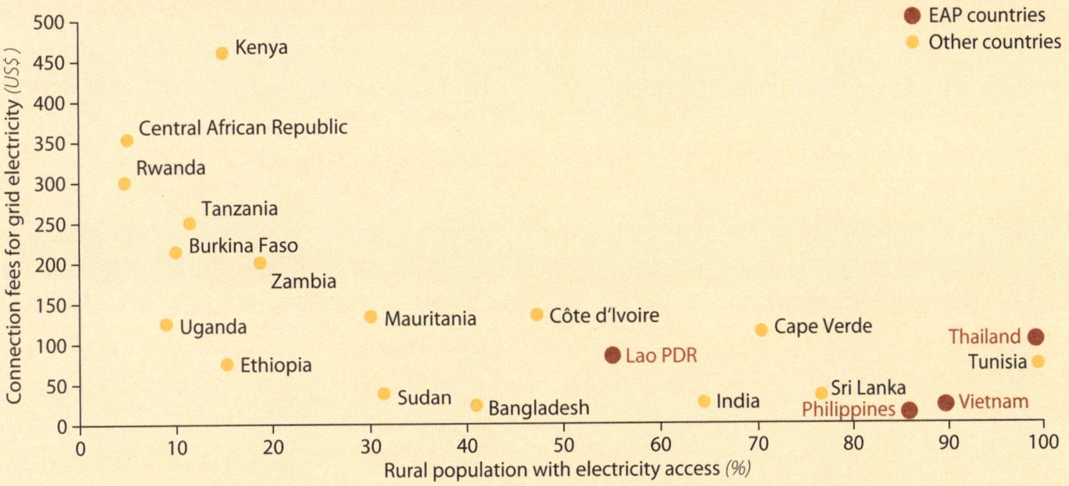

For most EAP countries, the connection fees are relatively low (box figure above). In Vietnam, the low connection fees (US$10) charged to a HH under the World-Bank-financed Rural Electricity Project I was made possible by distributing the real costs of grid extension, service drop, and metering among the Electricity of Vietnam Group (EVN), the provincial governments, the communes, and the newly connected customers/households.

Source: Adapted from Golumbeanu and Barnes 2010.

effective in addressing a country's equity objectives. Three principles have been used successfully to implement subsidies to achieve long-term financial viability:

1. ***Tariff/subsidy policy that recognizes the full cost recovery of an efficient service.*** The utility should be rewarded with the subsidy payment on confirmation that the service provided has been of adequate quality and quantity. Output-based aid (OBA) mechanisms appear to be a good choice for implementing this principle (box 3.6).
2. ***Quality control mechanism to ensure that payments to service providers are effectively linked to compliance.*** This mechanism requires monitoring the quality of service to clients and defining and imposing penalties in cases of noncompliance. The objective is to keep the focus on customer service.
3. ***"Ring-fencing" the finances of operation, maintenance, and customer services.*** These should be separated completely from investment activities, which are supported through capital subsidies.

In Vietnam, to ensure the financial viability of electricity institutions, the government designed the electrification effort with cost-sharing arrangements at all levels (national, EVN/PC, province, district, and commune government agencies), along with customer contributions, retained depreciation from EVN, and international donor support.

In the 1990s, Indonesia's PLN was connecting rural consumers in the Java-Bali region at a record rate but was unable to make significant progress outside this region. One of the main reasons was that the revenue under the uniform national tariff from customers outside the Java-Bali region did not cover PLN's higher operating costs in those areas, and there was no subsidy mechanism to compensate PLN for this loss (appendix 1: Indonesia).

Securing Finance for Investments in Electricity Access

A sector-wide framework can attract and sustain donor engagement and support. In low-income EAP countries, donor support through long-term financing is needed to supplement domestic sources of finance for subsidies that are designed for capital and operating costs of electrification projects. The financing gap in such low-income countries can be very large, making the alignment of donors' financial management and procurement procedures all the more important. In this regard, a spatial electrification rollout plan is of vital importance. Such a plan can be leveraged to develop a common financing platform for donors and external sources of concessional finance. The spatial plan can help present a clear and viable financing framework for the donors to coordinate their efforts and schedule assistance in step with the rollout.

In Thailand, the Provincial Electricity Authority (PEA) was able to secure low-cost capital and long-term loans for system expansion. Most loans from bilateral and multilateral sources in the 1970s and 1980s were long term and carried below-market interest rates; in some cases, no interest was charged (Barnes 2007). The loans significantly reduced costs and enabled PEA to build its revenue base before the loan repayment period began. In Vietnam, EVN financed the majority of its access investment needs through internal generation of funds. EVN's strong self-financing performance helped to leverage multiple sources of finance that were crucial to the program.

Leveraging climate change funds for off-grid electrification. Several private carbon funds provide carbon credits to off-grid energy projects that are using renewable energy resources. International agencies involved in the Clean Development Mechanism (CDM), particularly the World Bank's Community Development Carbon Fund, have supported solar home systems and recently expanded their interest to microhydro development and other renewable energy technologies.[21] However, small programs face significant obstacles in obtaining carbon funding, so it may be necessary to streamline procedures while observing CDM principles and methodologies. With greater facilitation, the many groups that are developing financing for off-grid renewable energy systems could achieve the dual goals of alleviating poverty and reducing carbon emissions. A rapid increase in the number of public and private "green funds" that are interested in supporting investments in renewable energy represents a new opportunity to attract additional support to renewable-energy-based electrification in EAP countries.

Consumer Focus and Timing of Electrification

Each EAP country has its own vision of providing electricity access to its population within a targeted timeframe. Overall, most countries are oriented toward achieving universal access to electricity by 2020–30 (table 2.2). In moving toward their access targets in a timely manner, countries will need to systematically prioritize the areas that are to be connected. Efforts to reach the HH who are forever inaccessible to the grid—the "permanent" off-grid areas—can and should proceed

21. Community Development Carbon Fund, World Bank, www.go.worldbank.org/QNLHGWLPS0

Table 2.2 Current National Targets for Electricity Access

Country	Population with electricity access (%)	Year
Cambodia	60	2020
Indonesia	90	2020
Lao PDR	90	2020
Mongolia	100	2020
Philippines	90	2017
Timor-Leste	80	2025

Sources: WHO and UNDP 2009; MIME 2009.

in parallel with grid expansion efforts. For those who live in areas that the grid will take 10–15 years to reach—the "transitional" off-grid areas—interim off-grid solutions are an option (figure 2.4). In addition, through innovative use of low-cost solutions, the expansion of the grid can be accelerated and the need for transitional solutions minimized.

To date, the experience of most countries is that their implementing capacity for off-grid deployment—given its nature and constraints—is far below that of the traditional grid rollout. Although the framework and capacity for off-grid service are improving, the proven implementation capabilities and speed of grid rollout still may seem attractive. All of these factors create a constantly shifting interface between the grid and off-grid space as the access effort gets underway. This constant flux calls for a dynamic spatial rollout plan for the national electrification effort.

The planning process can establish a clear system for prioritizing the areas to be electrified, while taking into account the conditions necessary for rural development: access to education and health services, an adequate transport system, agricultural potential, and access to markets. For example, Thailand employed a systematic and highly successful approach to plan for rural electrification through setting transparent socioeconomic criteria for prioritizing target areas for access expansion; integrating RE in the broader national development strategy; and taking into account equity considerations. Lao PDR and Vietnam followed similar approaches.

A spatial plan integrates the least-cost considerations of grid and off-grid options. As the grid spreads out and becomes gradually more expensive, off-grid options may become cost effective. Off-grid electrification may be the only option in places in which it is difficult to provide access such as in remote and mountainous areas and small, dispersed islands, all of which exist in several EAP countries. These areas can be considered "permanent off-grid areas" (figure 2.4). In areas in which grid extension is feasible but may not occur in the medium to long term (10 years), off-grid electrification is an intermediate option. Such areas may be termed "transitional off-grid areas," which gradually will shrink as the grid expands to cover them. In most countries, isolated generation and distribution facilities run by the informal sector had provided electricity services long before the national electrification programs reached these areas. In a systematically planned expansion, areas that are not likely to be reached by the grid in the near future can be served by off-grid facilities within a framework that would enable them to be absorbed subsequently into the grid.

Figure 2.4 Transitional Off-Grid Areas Shrink as Grid Expands

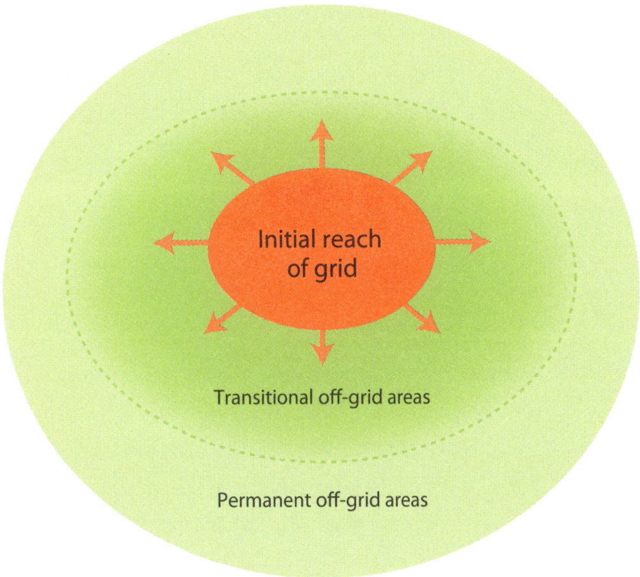

Source: Authors.

Box 2.5 GIS-Based Spatial Planning Platform: Powerful Tool for National Electrification Program Rollouts

GIS-based platforms have proved powerful planning tools for grid/off-grid national rollouts to meet time-bound targets. They also have been effectively used by the governments to syndicate large-scale financing on a programmatic basis. In Rwanda, based on a fully developed spatial plan, donors committed US$250 million over 5 years. Box figure 2.5.1 illustrates spatial planning in Rwanda for extending the electricity grid over 10 years. The box figure shows the "footprint" of the areas in which grid intensification will deliver the connection targets indicated in each planned state of the grid for 2009, 2010, 2012, and 2020. This prioritization has been set to achieve some grid electricity supply to 100 percent of sectors by the end of 2013. All social institutions falling within the shaded planning cells are earmarked for grid connection. The remaining social institutions targeted for electricity supply are shown by distinct symbols in the box figure and will be equipped with solar PV units. These results can be achieved by progressively extending the medium voltage network and by concentrating initially on increasing the number of connections within the areas already reached by the MV network. Given the high population density in the country, the plan shows that most areas can expect to be connected to the national grid. However, in some areas, local minigrids based on microhydro and solar PV systems will continue to be efficient for some time.

Box figure 2.5.1 GIS mapping for electrification rollout in Rwanda, 2009–20

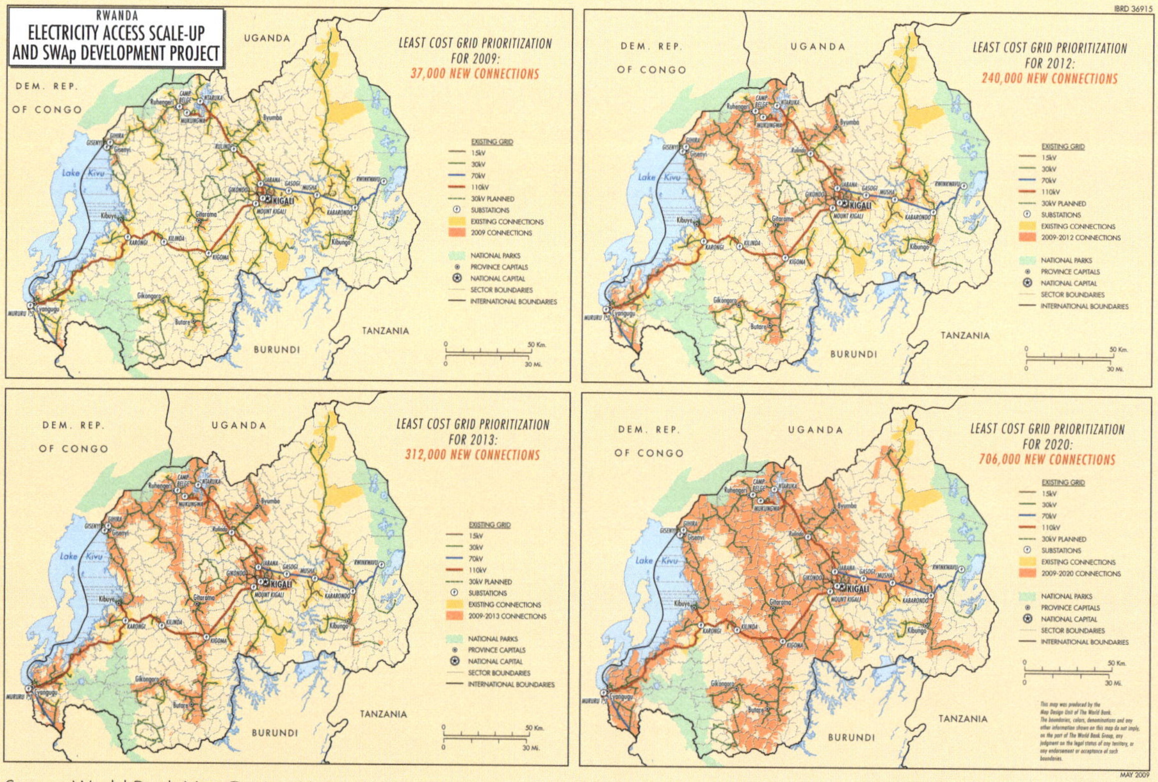

Source: World Bank Map Design Unit.

These transitional off-grid areas are best suited for renewable minigrid technologies (figure 2.4).

In contrast to the traditional and relatively static Master Plan analysis, a spatial plan based on GIS (geographic information system) models presents a far superior dynamic basis for systematic planning of the grid/off-grid interface. It enables the identification of communities and HH who are to be provided access by taking into account factors such as the existing infrastructure, population densities, and distance from the electricity grid, as well as the local renewable energy resources. The spatial model can be used to rapidly estimate and compare connection costs among different regions and communities. Other inputs for the model include electricity demand, costs, and geographic characteristics. The spatial nature of the model permits accurate representation of the existing electricity network and population distribution, which can form the basis for future expansion decisions. Box 2.5 illustrates an application of a spatial planning model for electrification and points out its additional advantage of attracting external funding for specific programs.

Universal and Business-as-Usual Electricity Access Scenarios

The current trends in electricity access in EAP countries (figure 2.5) are not likely to result in universal electricity access in the Region by 2030. This delayed timetable is not acceptable in the context of the Region's dynamic social and economic development, which is expected to continue in the coming two decades. This report has developed an alternative scenario that goes

Figure 2.5 Electricity Access: Estimated Results of Business-as-Usual and Universal Access Scenarios, 2010–30

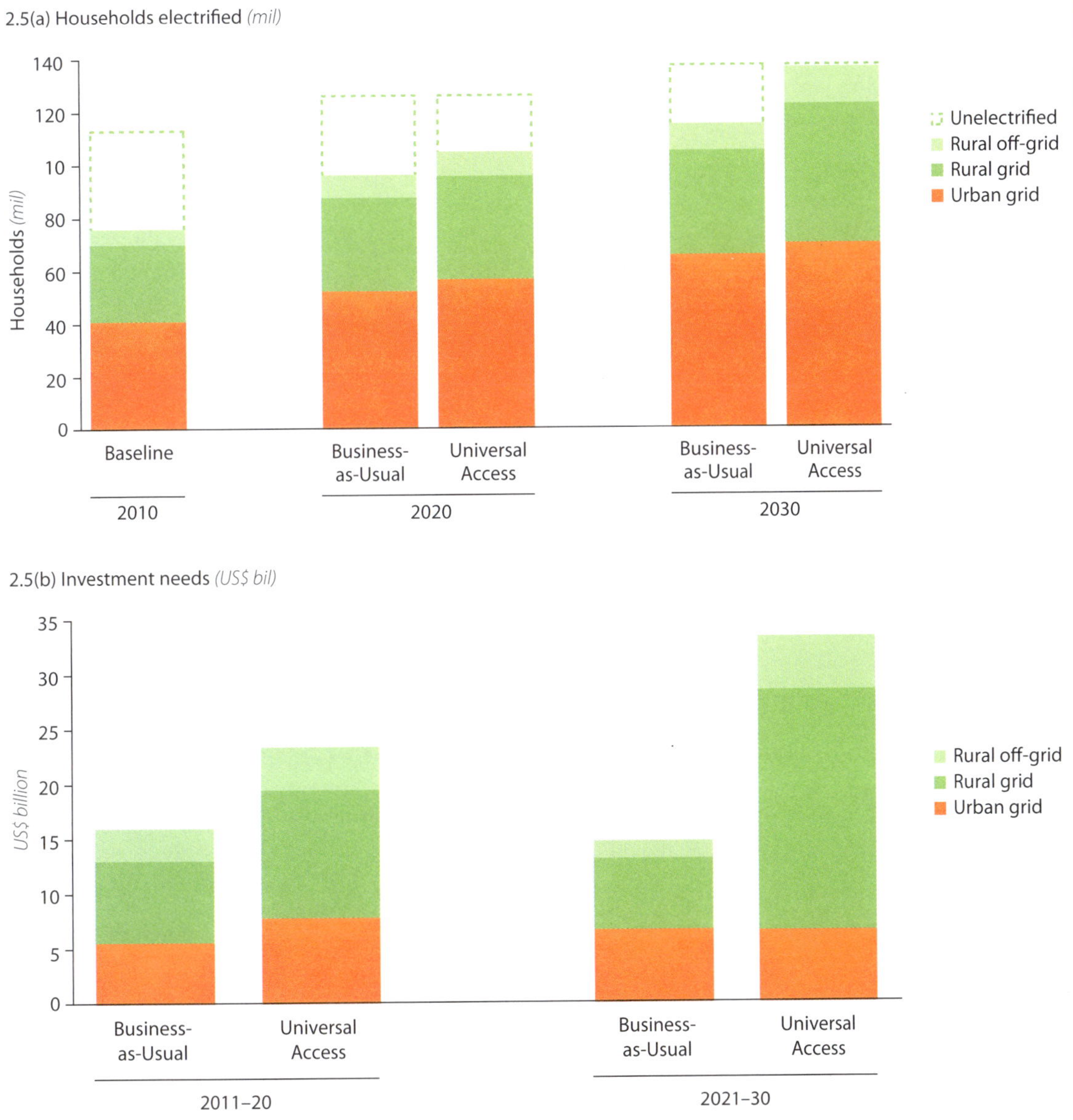

Source: Authors' calculations.

beyond Business-as-Usual and seeks to achieve universal electricity access in all EAP countries by 2030.

This report examines these 2 scenarios of electricity access growth up to the year 2030, and posits an intermediate result for 2020. These scenarios are not intended as forecasts or plans. Their purpose is to enable policymakers to understand the scale of investments needed, the technology options involved, and the feasible timeframes within which universal electricity access can be achieved.

- *The **Business-as-Usual scenario*** takes the average annual growth of energy access over the past decade (2000–09) as the "reference." It then assumes that the "reference" rate of growth will continue over the next two decades.
- *The **Universal Access scenario*** assumes that all countries will reach (near) universal access of 99 percent or more by 2030.

The two scenarios use standard projections for population growth and urban-rural population shares. Based on experience and expert estimates, these scenarios also make specific assumptions about the technology mix (urban grid, rural grid, and off-grid), average costs of connection for the urban and rural grids at different stages, and costs for off-grid applications.

Investment Requirements in the Business-as-Usual and Universal Access Scenarios

The scenario analysis shows that providing universal electricity access in EAP by 2030 would require (a) the electrification of approximately 60 million HH and (b) an investment of approximately US$56 billion. Under the Business-as-Usual scenario, the number of HH added would be approximately 39 million, and the required investments would be approximately US$30 billion. The Business-as-Usual scenario still would leave approximately 21 million HH without electricity access by 2030 (figure 2.5; tables 2.3, 2.4).

Table 2.3 Households Obtaining Electricity Access in EAP by 2030: Universal Access versus Business-as-Usual Scenarios

	Period	Total no. HH electrified in Business-as-Usual scenario	Total no. HH electrified in Universal Access scenario	Incremental no. HH in Universal Access scenario	Annual incremental no. HH in Universal Access scenario
No. HH connected *(mil)*	2011–20	19.8	28.6	8.8	0.9
	2021–30	18.8	31.2	12.4	1.2
	Total	**38.5**	**59.8**	**21.3**	—

Source: Authors' calculations.
Note: Any discrepancies in the totals in tables are due to rounding.

Table 2.4 Investment Needs for Electricity Access in EAP by 2030: Universal Access versus Business-as-Usual Scenarios

	Time	Total investment needs: Business-as-Usual scenario 2030	Total investment needs: Universal Access scenario 2030	Incremental needs for Universal Access	Annual incremental needs for Universal Access
Total est. cost *(US$ bil)*	2011–20	15.5	23.1	7.6	0.8
	2021–30	14.7	33.3	18.6	1.9
	Total	**30.2**	**56.4**	**26.2**	—

Sources: Authors' calculations.

Table 2.5 Investment Needs for Business-as-Usual and Universal Access Scenarios for Electricity Access in EAP Countries by 2030

	Total investment needs		Incremental needs for universal scenario			
Country	**Business-as-Usual scenario**	**Universal Access scenario**	**Total**	**Urban grid**	**Rural grid**	**Rural off-grid**
Low-access countries						
Cambodia	0.67	3.78	3.11	0.05	2.43	0.63
Myanmar	0.79	9.58	8.79	1.89	6.43	0.55
PICs (other)	0.05	0.35	0.30	0.04	0.03	0.23
PNG	0.12	2.54	2.42	0.12	1.72	**0.58**
Timor-Leste	0.06	0.36	0.30	0.03	0.10	0.18
Subtotal	**1.69**	**16.61**	**14.92**	**2.13**	**10.70**	**2.18**
Medium-access countries						
Indonesia	20.76	30.73	9.96	—	8.12	1.84
Lao PDR	0.69	0.69	—	—	—	—
Mongolia	0.13	0.13	—	—	—	—
Philippines	5.95	7.01	1.07	—	0.87	0.20
Subtotal	**27.53**	**38.56**	**11.03**	—	**8.99**	**2.04**

Source: Authors' calculations.

The incremental investment required for the Universal Access scenario over and above the Business-as-Usual scenario is US$26.2 billion. This requirement is divided in 2 tranches: US$7.6 billion in the first decade (2011–20), and US$18.6 billion in 2021–30. This division translates to an incremental requirement of US$0.8 billion per year for the Universal Access scenario in the first decade, and $1.9 billion per year in the second decade.

Incremental needs for low-access and medium-access countries. The total incremental investment needs for universal access (US$26 billion) are split between low-access and medium-access countries as US$15 billion and US$11 billion, respectively. Of these, the share of the rural grid is the largest at US$11 billion and US$9 billion, for low-access and medium-access countries, respectively. Regarding the incremental investment needs for the urban grid, low-access countries have an incremental requirement of US$2 billion. The medium-access countries do not have any incremental needs for their urban grids because this demand is fulfilled at the Business-as-Usual rate of growth. The share of the rural off-grid component is far lower than that of the rural grid. The shares for the rural off-grid component are the same for both low-access and medium-access countries at approximately US$2 billion (table 2.5).

Among individual countries, Indonesia needs the largest annual incremental investment at nearly US$500 million. Cambodia requires US$156 million per annum. However, this amount is a much larger share of Cambodia's current GDP (1.6 percent) compared to Indonesia's annual incremental investment, which amounts to 0.1 percent of GDP. At US$121 million, PNG's annual incremental investment requirement is similar to that of Cambodia (1.5 percent of GDP). The nominal incremental investment required for Timor-Leste appears small at US$15 million per annum, but it is a significant proportion of its current GDP at 2.7 percent. The Philippines requires a relatively small amount at US$53 million per annum, a negligible share of its current GDP (table 2.6).

Table 2.6 Universal Access Scenario for Electricity: Annual Incremental Needs by EAP Country by 2030 *(US$ mil)*

	Urban grid	Rural grid	Rural off-grid	Total	% GDP 2009 *(2009 US$)*
Low-access countries					
Cambodia	3	122	31	156	1.6
Myanmar	95	321	28	444	—
PICs (except PNG and Timor-Leste)	2	2	11	15	—
PNG	6	86	29	121	1.5
Timor-Leste	1	5	9	15	2.7
Subtotal	**107**	**536**	**108**	**747**	—
Medium-access countries					
Indonesia	—	406	92	498	0.1
Lao PDR	—	—	—	—	—
Mongolia	—	—	—	—	—
Philippines	—	44	10	53	Negligible
Subtotal	**0**	**450**	**102**	**551**	—

Source: Authors' calculations.

EAP countries must discover electrification solutions consistent with their geographies and natural resources, financial requirements, demographics, and social and political realities. Increasing electricity access is a dynamic problem-solving process. The nature of problems may change as programs evolve, but the underlying principles of cost-effectiveness, affordability, and timeliness of electricity service continue to guide successful programs. When cost recovery is pursued through sound system design and adequate attention is paid to its financial sustainability, most other program elements fall into place. Finally, service providers must take a customer focus by lowering barriers to electricity access and involving local communities in promoting electricity use.

Connecting houses in poor neighborhoods, Mongolia.

Photo credit: John D. Liu.

3

Electricity Access: Delivering Results on the Ground

As seen in earlier chapters, several EAP countries—China, Thailand, and Vietnam—have achieved near-universal access to electricity. In recent years, Lao PDR has made impressive strides toward that goal. Outside the Region, Brazil, Chile, South Africa, and Tunisia are notable examples of developing countries that have reached near-universal electricity in recent decades. In the past, Indonesia and the Philippines also led successful electrification programs, which now are in need of greater momentum. In terms of off-grid household electrification, Bangladesh and Sri Lanka have made significant progress in recent years (World Bank 2008b).

Each of these countries developed electrification programs suited to its own conditions, and no single model can be attributed to all of them. Nevertheless, all of these programs have demonstrated three common underlying principles: a sustained national vision and commitment to universal electrification, an enabling policy framework, and an accountable implementation system (figure 3.1). Following these principles will facilitate delivery of results on the ground in the EAP countries in which electricity access remains a challenge.

Sustained Government Commitment

Long-term national commitment is needed. Experience shows that it takes at least 1–2 decades to achieve universal electricity access. This timeframe is likely for the EAP countries with medium and low electricity access. A corresponding long-term national commitment over this entire period is essential to sustain the pace of electrification to its successful completion. This national commitment should be broad-based so that it is not unduly affected by the normal political and economic changes in a country.

In Tunisia, which reached near-universal access to electricity over a decade ago, the government had included electricity access as part of a broader, integrated rural development program that emphasized social equity. Chile's successful rural electrification program was founded on a high level of multiyear financial and political commitment through various national agencies including the energy sector regulator and the planning ministry, as well as through strong regional government buy-in (Barnes 2007).

Sustained government commitment was a cardinal feature of China's path to universal electricity access (box 1.3). In Thailand, RE was closely linked with the country's 5-year National Economic and Social Development Plans (Barnes 2007). Similarly, the Vietnam government's strong commitment to RE as a means to alleviate poverty and redress imbalances between rural and urban development was built into the government's 5-year plan and 10-year strategy (World Bank 2011b). Electrification is a top priority of

Figure 3.1 Principles Underlying Successful Electrification Programs

Source: Authors.

the Lao PDR government and an integral part of its growth and poverty reduction program. The government translated this commitment into time-bound targets for electricity access, and followed up with clear policy and financial commitment against the backdrop of its economy-wide policies for encouraging growth (World Bank forthcoming 2011).

In Cambodia, and some of the PICs (including Timor-Leste and PNG) that have low levels of electricity access, there is a clear need to establish a long-term national commitment. In Indonesia, the Philippines, and Mongolia, whose electricity access programs have slowed down, the governments must reaffirm their strong long-term commitments to universal electricity access, and take additional steps to create or revitalize the institutions that will translate the commitment into results on the ground.

Credible targets are necessary. Setting credible medium- and long-term targets in keeping with the potential of sector institutions is critical to achieve universal access to electricity. For example, in Lao PDR, the government set time-bound targets that consistently have been achieved by the national utility, Electricité du Laos (EDL).

From 13 percent at present, Cambodia's government has set a target of providing electricity access to 70 percent of rural HH by 2030. Under the country's current institutional and financial arrangements for electrification, this target may be difficult to achieve. The rural areas in the country are served by 249 small, privately owned rural electricity enterprises (REEs), most of which are in poor financial condition and lack sufficient technical expertise. Furthermore, the national power utility, EDC, which serves only the Phnom Penh area and some provincial centers, is not in a position to expand its activities due to high generation costs and low technical capacity. The credibility of the targets depends on the potential of the sector institutions to deliver results on the ground. If these institutions are financially or technically weak, they must be strengthened to put the targets within the country's reach.

Accountability is key. Indonesia has set a target of 95 percent–100 percent electrification of households by 2025. The Philippines has set a target of 100 percent electricity access by 2020. However, simply setting targets will not be enough. In general, the governments must be accountable to their constituencies, and the service providers must be accountable to their customers. Accountability also requires transparency so that results on the ground can be compared clearly against targets, and adjustments made as needed. In the Philippines and Indonesia, decentralization of governance to the provincial and district levels has brought in locally elected officials as key interlocutors. This development could be viewed as an opportunity to define, prioritize, and measure performance on the ground through intensive local engagement. Clearly, engaging the beneficiaries in all stages of the planning, implementation, and operational phases is critical.

Political interference should be minimized. In setting medium-term milestones, national policymakers should focus on setting objectives and measurable results indicators for the access program. Many electricity access programs in developing countries have failed to meet their targets due to significant political interference through lobbying by interest groups and politically connected communities. These pressures orient the implementing agencies more toward

Box 3.1 Thailand: Clear Criteria and Transparent Process for Electrification Expansion

Thailand used three principles to select villages for electrification:

1. Maximize potential benefits while keeping project costs reasonable
2. Integrate rural electrification into the broader national development strategy
3. Give consideration to the social and political requirements of less stable areas—a political imperative.

Within these boundaries, standard village selection criteria were applied. The selection process was carried out in three quantitative steps capped by a final qualitative step to determine which villages would receive electricity.

1. Village allocation for each province was based on socioeconomic conditions. This method favored the more advanced provinces, since the goal was to reach the largest number of beneficiaries within the project's budget.
2. Village allocation for districts within each province was based on the highest predicted levels of actual and potential demand for electricity, using the results of socioeconomic surveys conducted for this purpose.
3. Villages for final selection were based on three equally weighted factors: proximity of the grid and roads; village size and number of expected customers in the first five years; productive uses—agriculture, small industry and commercial establishments; and number of public infrastructural facilities (clinics, grade schools, potable water systems). The strategy of electrifying the more economically advanced areas first yielded returns on investment that contributed to achieving targets within 15 years instead of the planned 25 years.

Source: Barnes 2007.

meeting nominal political goals than toward providing timely professional service to their target beneficiaries.

In Thailand, the overall strategy was to electrify the more economically advanced areas first. Doing so would put the utility, PEA, on a sound financial footing and set the stage for further grid expansion. However, subsequently, it was recognized that social and political considerations required giving simultaneous priority to politically less stable areas. These two considerations were addressed through an objective process of goal setting and monitoring by PEA (box 3.1).

Lao PDR's rural electrification planning process takes into account the financial and economic viability of the investments. It also incorporates a simple but rigorous village screening and prioritization process to maximize social impact. Clinics, schools, temples, irrigation, and production activities get priority—within the available budget. The planning process features upstream consultations with provincial energy department officials and the relevant provincial authorities as to their priorities and regional development plans. Following a cross-cutting review of all provincial and branch proposals, a revised expansion plan is prepared and conveyed to branch offices for implementation (World Bank forthcoming 2011).

Vietnam's successful electrification program also established three clear principles for selecting the communes to be connected:

1. The commune should have reasonable potential for economic development to ensure that energy consumption will justify the investment.
2. Connecting the commune to the grid should be the least-cost solution to electrify it.
3. The selection should follow a process focused on economic sufficiency and participation of local communities and HH.

There were clear incentives for local companies to submit sound plans, which increased their chances of obtaining finance from provincial electricity companies. The public was kept informed of access expansion plans to keep the process transparent and to manage expectations.

Links with other sectors facilitate rural electrification. Meeting the national targets for electrification also will require that the government facilitate cross-sectoral support and links as necessary. Doing so may mean closer integration with other development programs; or it may mean that power sector agencies work closely with national, provincial, and local authorities. In Thailand, PEA actively sought and received strong local community support during all phases of system development; and developed a strong rapport with district, subdistrict, and village heads. As a result, PEA was able to obtain in-kind contributions at the time of construction for rights-of-way and the right to cut trees, with no financial compensation.

Outside the EAP Region, some governments have explicitly linked rural electricity access to other sectors. From 1993–2005, the Brazil Northeast Rural Poverty Alleviation Program used an integrated rural development approach under which approximately 650,000 families received electricity, and 802,000 families received water supply (World Bank 2010d). In Uganda, the Energy for Rural Transformation program explicitly features electrification of remote health clinics, schools, and water pumping via solar photovoltaic systems. The program also aims to provide electricity to rural agricultural and agro-based enterprises, such as honey processing, milk cooling, and maize processing (World Bank 2009).

The electricity provider's financial viability in rural areas improves if there are businesses that use electricity in the daytime. Household electricity use in rural areas is mainly in the evening hours (for lighting and TV); without daytime use, the system's capacity is lying idle most of the time. Hence, establishing links with other sectors that could promote productive, daytime uses of electricity would not only boost the productivity of the end-user enterprises but also improve the financial viability of the service provider.

Effective Enabling Policy Framework

The policy framework necessary to facilitate electricity access has four principal components: institutional, financial, technical, and regulatory. Most medium- and low-access EAP countries already have in place at least some elements of this framework and should be able to add or strengthen the other elements without much delay. However, some countries, such as Cambodia and several of the PICs, are likely to require external assistance for this purpose.

Institutional Framework

Designate an entity with clear responsibility for grid-based electrification. The government needs to designate or create one or more implementation agencies that will be responsible for achieving the national electrification targets through grid-based electrification. Medium- and low-access EAP countries already have national power utilities that operate the main grid. In most cases, such a utility is the most well-established power-sector entity in the country. It is charged with the responsibility to connect unserved urban and rural HH primarily through grid-based rural electrification.

Some countries have chosen to create separate entities or departments to focus on rural or off-grid electrification. For instance, early in its national electrification effort, Thailand created a new agency, the Office of Rural Electrification (ORE) under PEA for grid-based RE. Once the task of extending the grid to most villages had been completed, PEA dissolved ORE and resumed responsibility for serving villages and other nonurban provincial customers (Barnes 2007). Bangladesh has a separate Rural Electrification Board, which has shown encouraging results. Such options must be considered by each country in its own context.

Effective off-grid implementation agencies are needed. As noted above, off-grid electrification will play a significant role in the overall access effort. Thus, the roles of private sector, public-private partnerships, and community groups in off-grid electrification must be clearly specified. If multiple implementing agencies are involved in the off-grid electrification effort, there should be an institutional mechanism to coordinate their activities. Few EAP countries have well-defined entities for off-grid electrification. In the Philippines, some rural electric cooperatives (RECs) operate independent minigrids. However, the responsibility for electrification in "missionary" areas—those not connected to the main grid—is led by the Department of Energy, with a heavy reliance on the Missionary Electrification Development Plan (MEDP). This plan includes mechanisms by which areas unlikely to be served by the RECs in the near future are opened up for service provision by other means, including "qualified third parties." In Cambodia, the RECs are engaged in off-grid provision of electricity and are subject to regulation by the Electricity Authority of Cambodia (EAC). In other EAP countries, private entities are operating without a clear institutional framework. For instance, in the PICs, several dealers are engaged in selling "solar lanterns" with no clear institutional mechanism to coordinate their activities.

The experience of Latin America offers some insights that may be relevant to EAP. In Peru, qualified service providers undertake SHS electrification of targeted areas within an established technical and regulatory framework that covers equipment and service standards. The objective of these programs is the provision of electricity service. Beneficiaries pay a monthly tariff in a manner similar to customers of an electricity utility service. In Peru, any type of entity—private, public, or NGO—can bid to undertake off-grid electrification programs in specified areas. In several instances, the regional power utilities have won the bid to undertake these off-grid programs by leveraging their technical expertise and knowledge of the demographics and needs of the area. In Brazil, regional power utilities operate approximately 3,000 isolated diesel minigrid systems (Niez 2010). In Argentina, the regional electricity utility companies have set up subsidiaries whose core business is to manage and operate the off-grid SHS electrification service program (World Bank 2008a).

Coordination with local government and other agencies can greatly facilitate electricity access. Particularly in the rural areas, the service provider should coordinate with the local administrative units and agencies that can facilitate the electrification effort. These entities include the local government, which can help minimize delays and barriers by getting support and cooperation from various authorities; and environmental and social safeguard agencies, which must issue permits for the use of natural resources, such as water for minihydro schemes.

Productive uses of electricity should be facilitated. The relevant government agencies and service providers should coordinate with the local entities that are responsible for promoting income-generating activities in the rural areas. The use of electricity by local enterprises can increase their productivity and the overall level of economic growth in the area. Furthermore, the daytime use of electricity by enterprises that are able to pay more for electricity than poor HH can improve the financial viability of electricity service. During their years of expanding electricity access, the governments of China and Thailand deliberately promoted productive activities in the areas that were being electrified. China encouraged agroprocessing and other value-added activities, while Thailand built criteria in its rollout plan to cover areas with the promise of economic activity (Barnes 2007). There also should be a mechanism to coordinate the electrification program with agencies in end-user sectors such as health, education, and water, particularly for off-grid schemes to ensure that the needs of these sectors outside the grid are fully met.

Rural electrification programs can benefit greatly from the involvement of local communities. The participation of local communities from the start of a project offers the advantages of helping to improve its design and mobilizing cash or in-kind contributions. Community involvement also considerably increases local ownership, thereby helping to ensure the operational sustainability of the service.

In Thailand, PEA adopted a flexible, innovative, pragmatic approach in dealing with communities through paying due attention to customer service, service promotion, and marketing; and encouraging community involvement. PEA actively sought and received strong local community support during all phases of system development, beginning with the initial surveys and continuing through construction, electrification, operation, and maintenance. Local labor was hired when feasible. PEA was able to obtain in-kind contributions at the time of construction in terms of rights-of-way and the right to cut trees, with no financial compensation (Barnes 2007).

In Vietnam, under the service agent model created under the Rural Electrification Projects, local community members maintained LV systems on behalf of the Power Corporations (PCs), carried out simple repairs, and handled collections. Citizens' involvement helped ensure accountability within local communities, minimize nonpayment, reduce system losses, and significantly lower the costs of system operation and management (O&M) for the PCs (World Bank 2011b).

Table 3.1 Tunisia: Multiple Funding Sources for Rural Electrification

Budgetary allocation • Regional Development Program • Integrated Rural Development Program **Extrabudgetary funds** • Presidential Fund • National Solidarity Fund	**External debt** • African Development Bank (AfDB) • *World Bank* (WB) • French Development Agency (AFD) • Kuwait Fund **Off-grid program** • Credits from SPV suppliers (50%) • *World Bank* loan credits (25%) • NGOs (25%)

Source: World Bank 2005.

Financial Framework

Sustained funding is needed for upfront investments. Providing electricity access requires major upfront capital investment. State-owned power utilities in the medium- and low-access EAP countries do not have adequate equity funds or creditworthiness to undertake this investment. Whereas the private sector may be able to raise its own funds, it is not a major participant in electricity access programs in these EAP countries. Hence, it is essential that the government develop a financial framework to fund the upfront capital costs of electrification through a system of grants and loans.

Funds for capital investments in electricity access should be available on a stable and reliable basis, that is, the available funds should not fluctuate from year to year. Wide fluctuations would make it difficult for the implementing agencies to develop and implement effective medium-term plans. In countries with successful electrification programs, funds have been provided from both external and internal sources. Table 3.1 shows the diversified sources from which Tunisia financed its rural electrification program.

In Thailand, PEA received low-cost, long-term loans for system expansion. In the 1970s and 1980s, most loans from bilateral and multilateral sources were long term and carried below-market interest rates. In some cases, no interest was charged. These loans significantly reduced costs and enabled PEA to build its revenue base before the loan repayment period began (Barnes 2007). Brazil uses its Energy Development Account (Conta de Desenvolvimento Energético) for making grants to its "Lights for All" RE program. The grant funds come from payments for the use of public assets by private operators, fines collected by the regulatory agency from concessionaires and permit-holders, and annual quotas paid by electricity sellers (Niez 2010).

In Vietnam, the RE program has involved a major public investment effort, matched by significant local contributions. The power utility, EVN, was able to self-finance a significant amount of the capital costs necessary for RE; and in the early years, local governments and consumers also

Box 3.2 Cambodia: Proposed Sector-Wide Approach for Electricity Access Expansion

Over the next 20 years, due to Cambodia's low electrification rate and lack of adequate generation and transmission capacity, the investment requirements for the country's national electrification program are enormous. Currently, the main source of energy consumed in Cambodia is expensive diesel fuel. The country is building several hydropower plants and a coal-fired plant, which are expected to reduce the share of oil-fired generation, and thus reduce average power costs. A national transmission network is being developed, taking the form of a number of high voltage (HV) transmission lines connecting to the neighboring countries of Vietnam and Thailand, to be followed later by cross-border connections to Lao PDR. This network will enable the import of power at lower prices. Over time, these lines are expected to be extended to create a single interconnected national grid.

These ambitious plans will require Cambodia to mobilize significant ongoing financing. This mobilization can be done most effectively within a sector-wide, programmatic framework, rather than a fragmented, project-by-project and donor-by-donor approach. Led by the government of Cambodia, the sector-wide approach (SWAp) would seek to rally donor partner engagement in alignment with a common sector-wide investment program and implementation and financing plan ("Sector-Wide Prospectus").

A key step in developing the prospectus is to build a strategic and credible grid rollout plan, including implementation, investment, and financing details for the grid and off-grid components (2012–30). The anchor for such a plan could be a GIS-based least-cost spatial grid rollout planning platform and model framework, which also determines the off-grid complement (box 2.5).

Source: Authors.

contributed significantly. The government was able to attract donor funds to support its well-implemented RE program.

In the 1990s, Indonesia's successful RE program was financed by government and donor funds. However, the macroeconomic crisis that began in 1997 badly damaged the financial health of the power company and made it difficult for the government to provide funds. Meanwhile, donor funds for RE dried up. This lack of funds was the major reason for the decline of Indonesia's RE program, which has not yet regained its former rate of expansion (appendix 1: Indonesia).

To better coordinate and plan for the long term, a sector-wide approach for donor funds should be followed. Expanding electricity access is a capital-intensive program and requires significant long-term finance. Even in countries with deep financial markets, the appetite to support these initiatives through private capital flows is limited. Therefore, most EAP countries would benefit from donor funds to finance their electrification programs. In raising future donor funds, EAP countries would do well to move toward a sector-wide approach (SWAp) and programmatic framework. These purposes could be accomplished through formulating a "Sector-Wide Prospectus," which would consist of a medium-term investment program and a credible implementation and financing plan. Donors could choose to finance parts of this prospectus, instead of providing funds on a fragmented project-by-project and donor-by-donor basis (box 3.2).

Use climate change funds more aggressively. In recent years, climate change funds have become available to support renewable-energy-based electrification programs. Many private carbon funds are actively providing carbon credits to off-grid energy projects. International agencies involved in the Clean Development Mechanism (CDM), particularly the World Bank's Community Development Carbon Fund, have

Box 3.3 Raising and Channeling Subsidies for Rural Electrification

Successful rural electrification programs have been able to provide the investment and subsidy funds necessary to serve rural consumers in three ways:

1. ***Direct budget transfers.*** Some countries have transferred budgetary funds to subsidize the expansion of rural electrification. The Tunisian government annually assessed the subsidies necessary to provide electricity to new customers. Over time, as costs per new customer rose, the government increased the level of subsidy per HH. In this way, the electricity price could be kept at reasonable levels for rural consumers, without raising rates for urban consumers.
2. ***Cross-subsidies.*** Thailand and Vietnam relied on within-sector cross-subsidies for rural consumers to provide most of the financing for rural electrification. This method provided a stable and consistent source of funds.
3. ***Universal electricity charge and rural electricity fund.*** A third way to finance rural electrification is to place a small surcharge on existing electricity users that goes into a rural electrification fund (REF). This approach has been used in countries that have many electricity distribution companies, such as Brazil, Chile, and Peru. In the EAP Region, the Philippines has established a Universal Charge for Missionary Electrification (UC-ME), which is imposed on all electricity end-users as stipulated in the country's Electric Power Industry Reform Act (EPIRA). Cambodia has established an REF to support rural electrification projects, particularly off-grid schemes. It is expected that an REF will be less vulnerable to political interference because it protected by its own governance rules.

Sources: Barnes 2007; Republic of the Philippines 2009.

supported solar home systems, renewable energy power generation, and even the provision of energy-efficient light bulbs. While considerable experience with large-scale projects exists, a government may have to take some steps to enable small programs to overcome the burdensome international requirements of qualifying for such funds.

Well-designed subsidies are necessary to ensure affordability and the financial viability of the service provider. It is considered good practice to subsidize capital costs only and let users pay for at least the operational costs. However, the financial framework must design subsidy schemes that take consumer affordability and utility costs into account. To be effective in increasing access, the subsidy schemes should be well targeted. The total subsidies that are needed for low-income and low-consumption groups can be seen as a public service obligation burden that must be borne either by the budget or by within-sector cross-subsidy schemes (box 3.3). Care must be taken to ensure that the public service obligation burden does not become onerous.

Poorly designed subsidy schemes can undermine the financial health of power utilities. If the service provider is not compensated for the cost of connecting and serving poor HH, there will be little incentive for the provider to fulfill its role.

One approach to ensure the full cost recovery of an efficient service is to calculate the efficient costs of service provision individually for each specific utility or regional cost center. This approach makes it possible to determine the revenues that each utility should be allowed to collect through tariffs and other sources. This individual computation is completely compatible with the application of a nationwide uniform tariff system (same tariff structure and rates in each category). In this case, the regulator or relevant agency must design and implement a special-purpose mechanism for revenue compensation among distribution companies. This option addresses the fact that companies serving high-density urban zones with generally higher income consumers have lower costs than utilities that serve predominantly low-income customers in rural areas.

For example, an important element of the Peruvian model is the full recovery of costs to the concessionaire/distribution company. Costs of operation, maintenance, and provision of customer services are higher in rural areas, in which consumption levels are lower. The average monthly consumption per Peruvian customer in rural consolidated areas is 30kWh, and for a newly connected customer is approximately 12 kWh; compared to the average urban consumption of 100 kWh. The ability of rural and newly connected customers to pay for the service also is lower. Under these circumstances, extending access in rural areas is served by a system of subsidies that acknowledges the lower income and lower electricity consumption levels of rural households and the higher costs of supply resulting from the remoteness of the location.

Upfront connection charges should be affordable for poor customers. Upfront connection fees present a major barrier for poor HH in connecting to electricity. Many poor HH do not have an electricity connection even several years after their area has been electrified, because they cannot afford the connection fees. When the objective is universal access, poor consumers will require assistance in paying the connection fees.

Common approaches to subsidizing connection fees—which are considered to be capital costs—range from a delayed monthly payment of the connection fee over a relatively long period to a complete subsidy. Lao PDR and Vietnam have successfully used low connection fees to facilitate electricity access. A study conducted in Lao PDR found that approximately 20 percent of HH in a village remained unconnected even 10 years after the village was electrified because the households could not pay the upfront connection charge of approximately US$80–$100 (World Bank forthcoming 2011). In response, Lao PDR piloted the "Power to the Poor" (P2P) program to enable the poorest rural HH—typically headed by women—to pay the connection costs in installments (box 3.4). The results were encouraging.

Technical Framework

Establish appropriate technical codes for grid expansion in urban and rural areas. In setting up their networks, all power utilities must follow a technical code. These codes generally are designed with urban conditions in mind so they may be too stringent for rural expansion of the main grid, or for independent grids in isolated areas. In China, for example, costs were kept under control by developing two compatible national technical codes: one for high-demand regions; the other for poor and remote regions, with provision for a smooth transition to the more stringent code with future growth.

In contrast, early in its rural electrification program, Vietnam had to deal with distribution systems of low quality that used different distribution standards. The varying standards led to large system losses and added to costs. To respond, after broad consultation with both provincial and local companies, the national utility, EVN, established common technical standards so that the power companies could construct lines all the way to the HH using the same standard.

Clearly, each country must consider this issue on its own, and formulate the code(s) and standards based on the conditions prevalent in that country.

Facilitate introduction of technical innovations. The technical codes/standards for grid-based RE must be flexible enough to facilitate cost-reducing innovations.

As discussed in chapter 2, many countries have reduced construction costs by using technical standards that are adapted to rural/low demand patterns, such as single-phase supply and single-wire earth return (SWER). Shield wire technology has been adopted by several countries with highly significant savings in investment costs. To control costs, Brazil mandates the use of single-phase transmission for two-thirds of the rollout, subject to satisfying quality parameters. In recent years, Lao PDR has employed SWER and also deployed shield wire technology in the mountainous northern region, in which it can be expensive to use normal reticulation design and standards.

Box 3.4 Power to the Poor (P2P) Program: Affordable Connection Fees for the Poorest in Lao PDR

The "Power to the Poor" (P2P) Program in Lao PDR is a subsidized financing mechanism being implemented by Electricité du Laos (EdL) to provide affordable connection and indoor wiring to poor households. Designed with a gender focus, the P2P program enables the poorest rural households who cannot pay the entirety of these costs upfront to access the main electricity grid for basic services. The program's objective is to raise HH connection rates to 85 percent–90 percent in village communities who already are connected to the grid.

To minimize upfront payments, the program design utilizes participatory methods and gender-sensitive eligibility criteria. It particularly targets female-headed poor households. It provides eligible households with a no-cost "basic" 3/9 Ampere meter (low voltage) sufficient to enable an average household to use 2 light bulbs and a small electrical appliance such as a radio.

Eligible households initially pay an average of at least 200,000 Kips (approximately $20) and can receive an interest-free credit of up to 700,000 Kips (US$80) to cover the costs of installation and indoor wiring. A household repays the interest-free credit in equal installments of 20,000 Kips (US$2.50) to EdL over 3 years as part of the household's monthly electricity consumption bill.

Once connected, a typical beneficiary household's monthly repayment is US$1–$2, on top of electricity consumption charges of US$1–$3 per month. Without the subsidies and the connection, these HH typically would spend approximately US$3–$5 per month for vastly inferior traditional energy sources such as batteries, diesel lamps, and candles. The monthly savings on their energy expenditures are projected to be sufficient to enable households to fully repay their connection costs in three years.

Implementation of the P2P pilot phase resulted in 537 newly electrified households, 68 of whom were female headed. This was an overall connection rate increase from 78 percent to 95 percent between 2008 and 2009. Based on the promising results of the pilot phase, since March 2009, EdL has scaled up the P2P program to cover approximately 8,000 additional households nationwide within 3 years.

Source: World Bank forthcoming 2011.

EDL has adapted innovations that help to lower investment costs of the national electrification program. An example is the low-cost concrete poles based on simpler production processes that make it feasible to produce the poles in "mobile factories" that are easily dismantled and moved to follow the progress of construction as the network spreads farther out. These mobile factories are a significant achievement as pole costs typically represent approximately 33 percent of the total investment costs of rural electrification programs.

Flexible technical standards also are needed for off-grid systems. Frequently, no well-defined technical standards govern off-grid systems—whether minigrids or standalone HH systems. Without suitable technical standards to govern their development, ensuring the quality of the service provided by these systems will be difficult. Developing these standards need not be a time-consuming task because they have been used successfully in many countries. This international history forms a good starting point from which to develop local standards.

It is important that the technical standards for off-grid systems are not so rigid as to make it difficult to introduce new technologies/products that either reduce costs or provide better service. Instead, what is needed is the technical capacity to test and certify these new technologies/products. One example of these emerging technologies is

Box 3.5 New Options for Serving Basic Electricity Needs in Off-Grid Areas

A range of new small low-cost devices for charging the batteries of efficient LED lights, mobile phones, and small radios is becoming available in all regions of the world. These user-friendly devices offer the prospect of quickly and cheaply improving the quality of life in off-grid areas. Such lighting solutions can cover room lighting, lights to illuminate a defined working area, and portable torch lighting. Furthermore, a new generation of solar lanterns, often called microenergy household solutions, is becoming available that can possibly operate a radio (perhaps a TV in the future). The variety of systems that is available in the market is growing, and prices are becoming more affordable, making them attractive interim solutions for households awaiting full electrification.

For example, in Vanuatu, retail prices for such products are approximately US$30–$60, depending on whether the device provides lighting only, or lighting plus charging for mobile phones and radio batteries. The systems have no operational costs except battery replacement. Rural mobile phone users in Vanuatu without electricity access pay at least US$1 per recharge in the outer islands, which means that their savings would quickly pay for the solar lanterns.

There are two key barriers to the rapid penetration of these devices. One is that there are no well-defined standards or certification schemes that give the consumers some assurance about the quality of the product. The second is that it is difficult for many HH to pay the upfront costs. The growing experience from the joint World Bank/IFC ongoing "Lighting Africa" initiative provides lessons in developing the institutional framework to support the systematic growth of applying these low-cost off-grid devices. Such framework should cover providing product quality assurance, market intelligence, consumer education, business support/access to finance, and policy and public sector operations (www.lightingafrica.org).

Box figure 3.5.1 Solar-powered LED light and cell phone charger

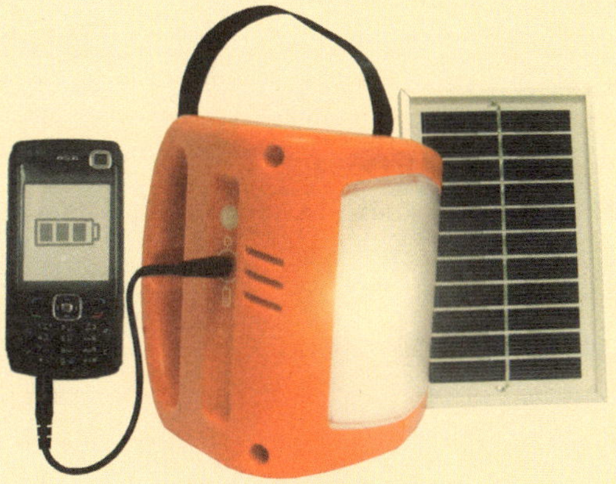

Photograph: http://www.dlightdesign.com

Sources: Aspden 2010; PPI 2009.

low-cost, solar-powered battery-charging systems, often called solar lanterns (box 3.5). These systems provide a basic amount of electricity. However, this amount of electricity is sufficient to quickly and inexpensively change the lives of people in remote areas who will not receive conventional electricity for many years and who cannot afford to buy larger solar home systems.

Regulatory Framework

Some EAP countries have chosen to set up an independent regulator for the power sector; others have managed the sector directly via a government department or agency. Regardless of which mode a country chooses, a well-functioning regulatory system is necessary to undertake key functions. Without such a regulatory system, the implementing agencies would find it difficult to meet their electrification targets.

Establish tariffs and cross-subsidy schemes that make the service provider financially viable. The regulator is responsible for setting tariffs and approving cross-subsidy schemes that transfer funds from higher-income users to finance system expansion. These tariffs and schemes must be designed with three competing pressures in mind: to ensure the financial viability of the implementing agencies, to make electricity access affordable to poorer people, and to keep the cross-subsidy charges imposed on higher income users reasonable.

Encourage private sector and community participation in off-grid schemes. Experience with off-grid schemes from several countries demonstrates that the potential exists for private sector participation, including public-private partnerships and community schemes. However, some preconditions need to be met, notably, a suitable regulatory framework, well-defined financing arrangements, technical assistance (TA), and capacity building support within which the private sector firms operate. In general, small private or community organizations operating in remote areas need "light-handed" regulation to keep the compliance costs manageable and similar to those of other microentrepreneurs.

Service providers must be made accountable to consumers. The ultimate aim of an electrification program is not only to extend electricity service but also to provide consumers with a quality of service that meets their needs and expectations. When actual or potential consumers do not find the electricity service up to their expectations, they must be able to convey their dissatisfaction to the regulator simply and inexpensively; and the regulator must be able to respond to these consumers. The regulatory system must provide for a suitable "consumer voice" mechanism that will help make the service providers accountable to the target beneficiaries for the quality and timeliness of the service provided to them.

Manage the grid and off-grid boundaries. The boundaries between the grid and off-grid options for rural electrification (chapter 2) can change as a result of changes in technologies, costs, and incomes. The regulatory system must manage the boundary on a rolling basis so that consumers in each area receive service from the appropriate provider with no overlap or confusion about which entity will provide service in a particular region.

Accountability for Results

Using output-based aid (OBA) to ensure accountability. Output-based aid approaches are being implemented in the energy sector in several Regions to improve access and targeting for the poor. Compared to input-based schemes, OBA shifts performance risk and accountability to service providers, because payments to them are made only after verified delivery of access and service. How much performance risk is borne by the OBA service provider depends on three components: the definition of outputs, the extent of phasing in subsidy payments, and the service provider's ability to prefinance the investments and services until subsidy payments are made (box 3.6).

Make service providers accountable to consumers. In the end, the results on the ground will be delivered by the designated implementing agencies. These agencies must be held accountable to the government and consumers. There must be a mechanism to monitor the progress of each implementation agency, and the results should be made transparent and available to the public. The public reporting system should cover the use of public funds by the implementing agency in achieving its results. The consumers should be able to discuss the progress of electrification and register their complaints with each service provider. They also should be able to convey their complaints to the regulator if the service provider does not meet expected standards. With the spread of cell phones and the internet, establishing a system of transparent reporting of service provider performance is feasible, as demonstrated in Indonesia and the Philippines in the environment and water sectors, respectively.[22]

---

22. Indonesia's PROPER (Program for Pollution Control, Evaluation, and Rating) is a national public environmental reporting initiative that has been disclosing data on industrial pollution of water bodies in metropolitan Jakarta for the past 15 years. In Manila, a public performance assessment system managed by the Metropolitan Waterworks and Sewerage System (MWSS) Regulatory Office has been reporting on

Box 3.6 Applying Output-Based Aid (OBA) to Improve Accountability in Providing Energy Access

The use of OBA is most widespread in individual off-grid systems for rural electrification. The "output" in an off-grid project is usually the installation of a functioning off-grid unit, such as a solar home system (SHS). Most OBA projects use the "dealer model," whereby private dealers sell systems in the open market. A third-party financial institution, such as a microfinance institution, may provide household credit for the initial down payment. Consumers own the standalone systems and, following the warranty period, are responsible for maintenance.

Typically, the "outputs" in grid-based systems are verified working connections to the network. An output also could be a specified period of service delivery, demonstrated through billing or collection records. The OBA subsidy generally is used to buy down the capital cost to make access affordable and is paid once installation is verified.

In addition to one-off capital subsidies for access, transitional and ongoing output-based subsidies have been used in grid-based schemes, such as the Pamir Private Power Project in Tajikistan. An OBA program in Ethiopia offers a $35 payment to the utility, Ethiopian Electric Power Corporation (EEPCo), upon verification by an independent agent of pre-agreed outputs, including the connection and sustainable provision of services. As a result, the number of electrified towns in Ethiopia has grown very fast during the last five years.

OBA schemes are only as sustainable as the regulatory environment in which they operate. They cannot bypass the need for a strong regulatory regime. Ultimately, an output-based capital investment program is only one part of a larger service delivery arrangement. To provide sustainable service over time, tariffs need to be set at appropriate levels, and subsidies need to be minimized. There also are additional costs associated with implementing OBA. Monitoring and verification require capacity and resources. Capacity can be an issue in OBA schemes, especially with small and local private providers.

Sources: Mumssen and others 2010; World Bank 2010a.

Monitor the extent of cost reductions achieved by the service providers. Given the high costs of electricity expansion and the associated subsidies, it is important that service providers make an effort to reduce costs. In the EAP Region, the utilities in Thailand and Lao PDR have shown that it is possible to develop a culture of cost-cutting and introducing low-cost innovations. Outside the EAP Region, South Africa and Tunisia also have cut their costs significantly. Hence, an element of the accountability of the service providers is their ability to reduce costs over time.

This chapter has covered three important principles for driving results on the ground to achieve electricity access: sustained government commitment, an enabling policy framework, and a focus on accountability for results. These principles have derived from the experience of countries within the EAP and other Regions that have succeeded in providing near-universal access to their populations. There are no straightforward recipes for applying these principles. Each EAP country must formulate policy solutions that are consistent with its electricity access status, its institutional and financial strengths and weaknesses, and its demographic and geographic realities.

---

the service quality for water and sanitation services for the two water concessionaires.

Traditional cooking methods rely heavily on the use of solid fuels and inefficient stoves.
Top: Charcoal market, Cambodia.
Bottom: Cooking on a three-stone stove, Cambodia.

Photo credits (top and bottom): Michael Wild.

4

Modern Cooking Solutions: Status and Challenges

The promotion of modern cooking solutions in EAP is a challenge that will require attention for at least the next 20 years. One the one hand, the promotion of modern cooking fuels—principally LPG and natural gas—is the responsibility of large energy companies that have difficulty in serving even the existing customer base in many EAP countries. Many such firms have not really begun to tackle the challenge of providing affordable modern fuels to poor sections of the population. On the other hand, the promotion of advanced cookstoves that can use biomass, charcoal, or coal cleanly and efficiently is not attractive to the private sector due to low affordability among poorer households, the main target group. In other words, it is not financially attractive to sell either modern fuels or advanced stoves to relatively poor populations.

The poorest sections of the population, who live primarily in rural communities form the "bottom-of-the-pyramid" opportunities. This population segment traditionally has been viewed as being relatively unattractive to investors. The economic and social benefits of reaching these markets with modern cooking fuels and advanced cookstoves are significant. The benefits include fewer premature deaths; improved health and productivity; hours of drudgery avoided, mostly for women; less pressure on the local ecosystem from wood collection; and perhaps even a reduction in greenhouse gases. International coalitions are giving increased attention to developing business models that promote modern cooking solutions for the poorest households (IEA 2010; UN Foundation 2010). In recent years, new options have emerged that may accelerate this access.

Relationship among Traditional Cooking Solutions, Health, and Poverty

The use of biomass and solid fuels for cooking is extremely common in EAP, even in countries with high rates of modern fuel use. This pervasive use of solid fuels—including wood, coal, straw, and dung—and traditional cookstoves[23] results in high levels of household air pollution, extensive daily drudgery required to collect fuels, and serious health impacts.

It is well known that open fires and primitive stoves are inefficient ways of converting energy into heat for cooking. The average amount of biomass cooking fuel used by a typical family can

23. The use of biomass energy in inefficient or open stoves is considered a traditional way of cooking. On the other hand, gas, LPG, kerosene, electricity, and biomass energy used in efficient or less polluting stoves are considered modern ways of cooking. Other examples could be used, but the general idea is that traditional ways of using energy typically are inefficient and somewhat polluting, whereas the opposite is true of modern energy use for cooking.

Table 4.1 Annual Premature Deaths Attributed to Air Pollution from Cooking with Solid Fuels in EAP Countries, 2007 *(mil)*

EAP Region	Pneumonia (children under 5 yrs)	Chronic obstructive lung disease (adults older than 30 yrs)	Cancer (adults older than 30 yrs)	Total premature deaths
Total	40,800	589,300	34,700	664,800

Source: WHO and UNDP 2009.
Note: According to the definitions used by WHO and UNDP 2009, solid fuels (SFU) refer to traditional biomass—wood, charcoal, dung, straw, and crop residues—and coal. The figures in this table were computed by WHO to ensure compatibility. Thus, they are not necessarily the official statistics of WHO member states, which may use alternative rigorous methods. The number of deaths attributable to indoor air pollution from solid fuel use was calculated without removing chronic obstructive pulmonary disease and lung cancer deaths, both attributable to smoking, thus leading to higher figures than previously reported. The percentages of population using SFU, which were used as an exposure measure to calculate the present death figures, were published in WHO 2006 and in the WHO Country Profiles of the environmental burden of disease (WHO and UNDP 2009). For global methodological reasons, WHO does not exclude smoking from its more recent statistics on the number of premature deaths due to solid fuel use. However, as noted above, not excluding smoking produces current figures that are higher than some previously published figures. For instance, excluding the number of deaths due to smoking would reduce the number of premature deaths due to chronic obstructive pulmonary disease by approximately 33 percent. Nevertheless, a very high number of premature deaths due to HH solid fuel use would remain.

be as high as two tons per year.[24] Indoor biomass cooking smoke also is associated with a number of diseases, including acute respiratory illnesses, cataracts, heart disease, and even cancer. Women and children in particular are exposed to indoor cooking smoke in the form of small particulates up to 20 times higher than the maximum recommended levels of the World Health Organization.[25] It is estimated that smoke from cooking fuels accounts for nearly 2 million premature deaths annually worldwide[26]—more than the deaths from malaria and tuberculosis combined.

In EAP, the number of deaths that can be attributed to cooking with coal and biomass fuels exceeds 600,000 per year (table 4.1). The Region accounts for approximately 33 percent of the world's deaths attributable to these diseases. Young children are especially vulnerable since they spend much time indoors close to their mothers, including while they are cooking. A meta-analysis of global studies on pneumonia risk in children under 5 years indicated that children exposed to smoke from solid fuels were over 1.8 times more likely to contract pneumonia than children in households who do not use solid fuels.[27] In addition, strong evidence supports the causal linkages between biomass combustion emissions and acute respiratory infection among children.[28] Thus, there also is significant evidence linking indoor air pollution to a variety of health problems prevalent in developing countries.

Biomass fuel often is collected from the local ecosystem, most often by women and children. This time-consuming drudgery diverts time from productive and family activities. Family members spend a considerable amount of their human energy collecting fuel, whether from common village land or farmers' fields. The time spent collecting fuel sometimes can be as high as one hour per day (World Bank 2002). Biomass fuel collection often entails walking long distances carrying heavy headloads and enduring safety hazards. Furthermore, it can lead to a gradual deterioration of the local environment and depletion of biomass supplies, meaning even longer walks and greater drudgery.

24. World Bank 2011a.
25. WHO 2005.
26. WHO and UNDP 2009.
27. Kammen and others 2002; Parikh and others 2001; Smith and others 2004: 1435–94.
28. Dherani and others 2008; Smith and others 2004.

Traditional Cooking Solutions and Poverty

Overall, the use of solid fuel and biomass is closely intertwined with poverty (figure 4.1). EAP countries conform to worldwide patterns in which the percentage of a country's population who use biomass for cooking is significantly related to the level of GDP per capita. Based on the cross-sectional data for 2007, the general pattern is that the use of biomass fuels declines as GDP per capita increases, although at a far slower rate (figure 4.1). The proportion of HH using biomass for cooking declines approximately 0.16 percent for every 1.0 percent of income growth.

In EAP, the highest percentage of people using modern fuels for cooking is in the relatively higher income urban communities of Indonesia, the Philippines, and Thailand. However, even in Thailand, whose LPG use is the highest in the Region, well over 33 percent of the population use solid fuels such as wood, straw, or charcoal as their main cooking fuels (WHO and UNDP 2009). Perhaps because coal is used for heating and cooking, the countries in the next group are China, Mongolia, and Vietnam. The remaining countries in the Region, which have lower GDP per capita, are heavily dependent on biomass fuels such as wood or agricultural residues for cooking. In this third group, which includes Cambodia, Lao PDR, and Myanmar, 70 percent–90 percent of the people use biomass energy for cooking.

Figure 4.1 Household Biomass Energy Use versus GDP per Capita in Developing Countries, 2007

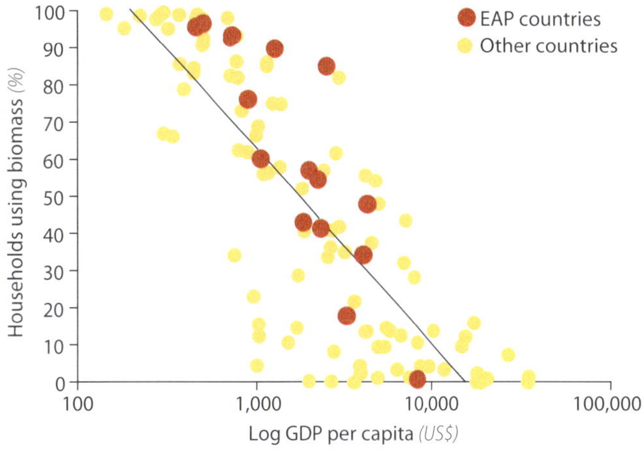

Source: WHO and UNDP 2009.

Gender Dimensions of Clean Cooking

Although women have cooked with traditional biomass stoves for millennia, it is only in recent years that the accompanying health risks and work burden, as outlined above, have been fully understood. The past "invisibility" of this issue to policymakers explains, in part, why governments and international organizations previously failed to assign this issue the priority it deserves. Much of the time and energy women spend on domestic tasks remains invisible to policymakers since non-market productive work is not counted in economic statistics or national accounts (box 4.1).

At present, in part because women are at the lowest level of paid and unpaid work, women's energy uses often are not considered a priority by policymakers, who are focused on high-profile energy projects. In formulating policies and investment priorities, energy policymakers should take account of women's non-market productive work. Clearly, the lives of women—especially poor women—and children stand to benefit significantly from energy programs that take into account the output produced by women and the burdens they bear in the production process. However, it must be acknowledged that the relationship between interventions and changed exposure to cooking smoke is complex. For instance, in China it was found that, in cold climate regions, the type of stove had more of an impact on exposure to pollution than the programs that attempted to change cooking behavior (World Bank 2007a). The conclusion was that, to have an impact on lower exposure to pollution levels, training in fire-tending techniques needs to be coupled with the use of different stoves.

Solutions for reducing the burdens on women from traditional cooking practices are likely to involve smaller, demand-side interventions and investments different from the more visible large-scale energy projects. The strategies for small-scale electricity goods and services such as solar home systems and lamps actually are very similar to those that involve the development and implementation of clean cooking solutions. Depending

Box 4.1 Women's "Invisible" Work

The language used to describe nonmarket productive work has contributed to the invisibility of women's time- and energy-intensive work. For example, cooking is considered "active labor" when cooked food is sold, but "economically inactive labor" when it is not. Similarly, housework is "productive" when performed by a paid domestic servant, but "nonproductive" when no payment is involved. Those who care for children at an orphanage are considered "occupied," while mothers who care for their children at home are "unoccupied." An electric pump that transports water is counted as part of the economy, whereas a woman who carries water is not. A water mill that grinds grain is "economic output," but a woman who performs the same task with mortar and pestle does not count. Trucks that consume fossil fuel to transport crops are considered part of GDP, while women who carry headloads of crops never show up in national accounts.

Source: Cecelski 1995.

on the possibilities within countries, it may be possible to merge these similar agendas to achieve the scale necessary to enable larger and more visible investments.

Status of Cooking Fuel Use in EAP

The widespread use of biomass energy is one reason that the quality of stoves used by households in developing countries is so critical. Approximately 50 percent of people in the East Asia and the Pacific Region use solid fuels—coal, wood, dung, and agricultural residue—as their main cooking fuels (table 4.2). Of these, close to one-half billion people rely on biomass energy for cooking, and for these populations, the collection of biomass is a frequent and very arduous task.

There are significant rural and urban differences in cooking fuel use in EAP. Modern fuels such as LPG and, to a lesser extent, electricity are fairly prevalent in urban areas. Kerosene, coal, and charcoal are transitional fuels because they generally are bought, not self-collected, and are more convenient to use than traditional biomass fuels. As might be expected, the number of people in rural areas cooking with solid fuels is higher than in urban areas. Although the health implications of cooking with these fuels are similar, most of the solid fuels burned in rural areas are not purchased but are collected from the local environment. The traditional fuels, comprising wood, straw, and dung, generally are the predominant type of cooking fuels in most rural areas in the Region (figure 4.2).

The countries that face the greatest challenges in moving to modern cooking systems are Cambodia, Lao PDR, and Mongolia in both urban and rural areas. Indonesia, the Philippines, and Vietnam show a greater use of kerosene and charcoal, but also significant traditional use of biomass and solid fuels in rural areas. Even though its share of traditional fuel use is lower, China has by far the largest absolute number of people who still do not cook with modern fuels such as LPG or electricity.

Despite rapid modernization in the Region, the transition to modern cooking fuels and clean and efficient stoves has been much slower than electrification. The number of people who depend on traditional biomass fuels and coal in the Region is still quite high (figure 4.3). Close to 1 billion people still use solid fuels for cooking and heating. Of these, approximately 40 percent, or approximately 400 million people, mainly in China[29] and Vietnam, use coal. The other 600 million are dependent on wood, straw, or animal dung for cooking. Thus, cooking and/or heating with solid fuels remains pervasive in East Asia and the Pacific.

29. WHO and UNDP 2009; Government of China 2008; Carolina Population Center and NINFS 2008. Numbers provided by industrial and commercial sources may differ from this estimate, which is based on household surveys.

Table 4.2 Population Relying on Traditional and Modern Fuels in Developing Countries, 2007 *(mil)*

	No. of people using solid fuels			People using modern fuels
Country group	Traditional biomass	Coal	Total	LPG, kerosene, gas, electricity
Total developing country	2,564	436	2,999	2,294
Less developed countries	703	12	715	74
East Asia and the Pacific	552	427	986	948

Source: WHO and UNDP 2009; authors' calculations.

With the exception of China, most other EAP countries have not focused on developing significantly scaled-up programs to propagate clean and efficient cooking. In sum, cooking with solid fuels on traditional stoves appears to be the rule rather than the exception in EAP. Today, promising alternatives to such traditional cooking practices exist (chapter 5).

Households in China and Mongolia use stoves for heating, and these stoves have problems similar to those of cookstoves. Many of the traditional heating stoves burn fuel very inefficiently, leading to higher fuel costs and extensive pollution, especially in periurban areas. Studies conducted in Northern China on the impact of improved stoves on household air pollution showed that new stoves and better ventilation techniques resulted in much higher efficiency and lower emissions than did the old stoves (World Bank 2006a; World Bank 2007a). The new stoves decreased fuel consumption by 30 percent–50 percent. Measurements taken under controlled conditions show a more substantial reduction in concentrations of particulate matter (PM) and carbon monoxide (CO) (13 percent–15 percent). Thus, in the northern provinces of China and in Mongolia, this issue of traditional heating stoves is extremely important and can be addressed in a way similar to the promotion of clean cooking.

Policies to Encourage Clean and Efficient Cooking

The extent of the adoption of modern fuels depends on three primary factors: (1) level of income, (2) price of the fuels, and (3) physical access to fuels (table 4.3). As their incomes rise, households in developing countries generally switch to LPG fuel and various types of specialized electric cooking appliances (Barnes and others 2005). However, even under optimistic projections of income growth, the use of biomass fuel among developing-country households will continue for years to come (IEA 2010). In

Figure 4.2 Rural-Urban Divide in Use of Modern Cooking Fuels in EAP Countries, 2008

Population not using modern cooking fuels, 2007 *(%)*

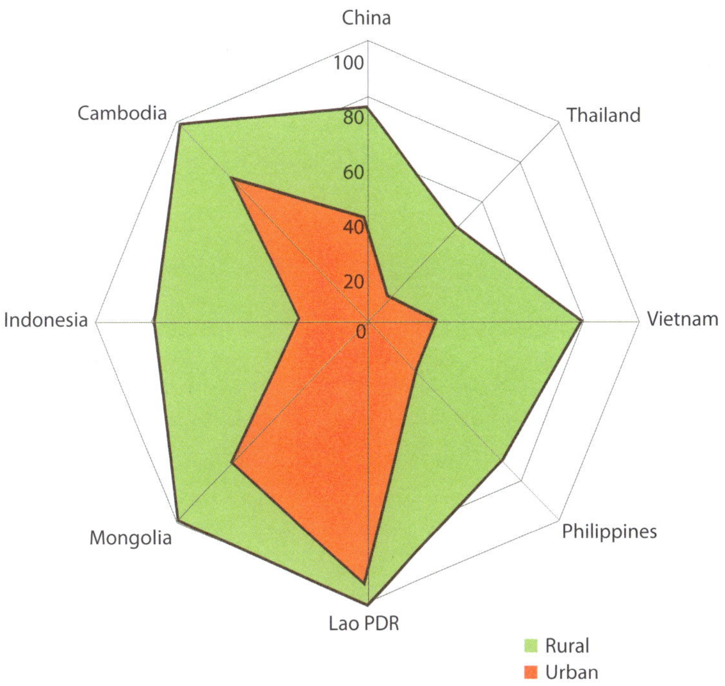

Sources: WHO and UNDP 2009; authors' calculations.

Figure 4.3 Patterns of Cooking Fuel Use in EAP Countries

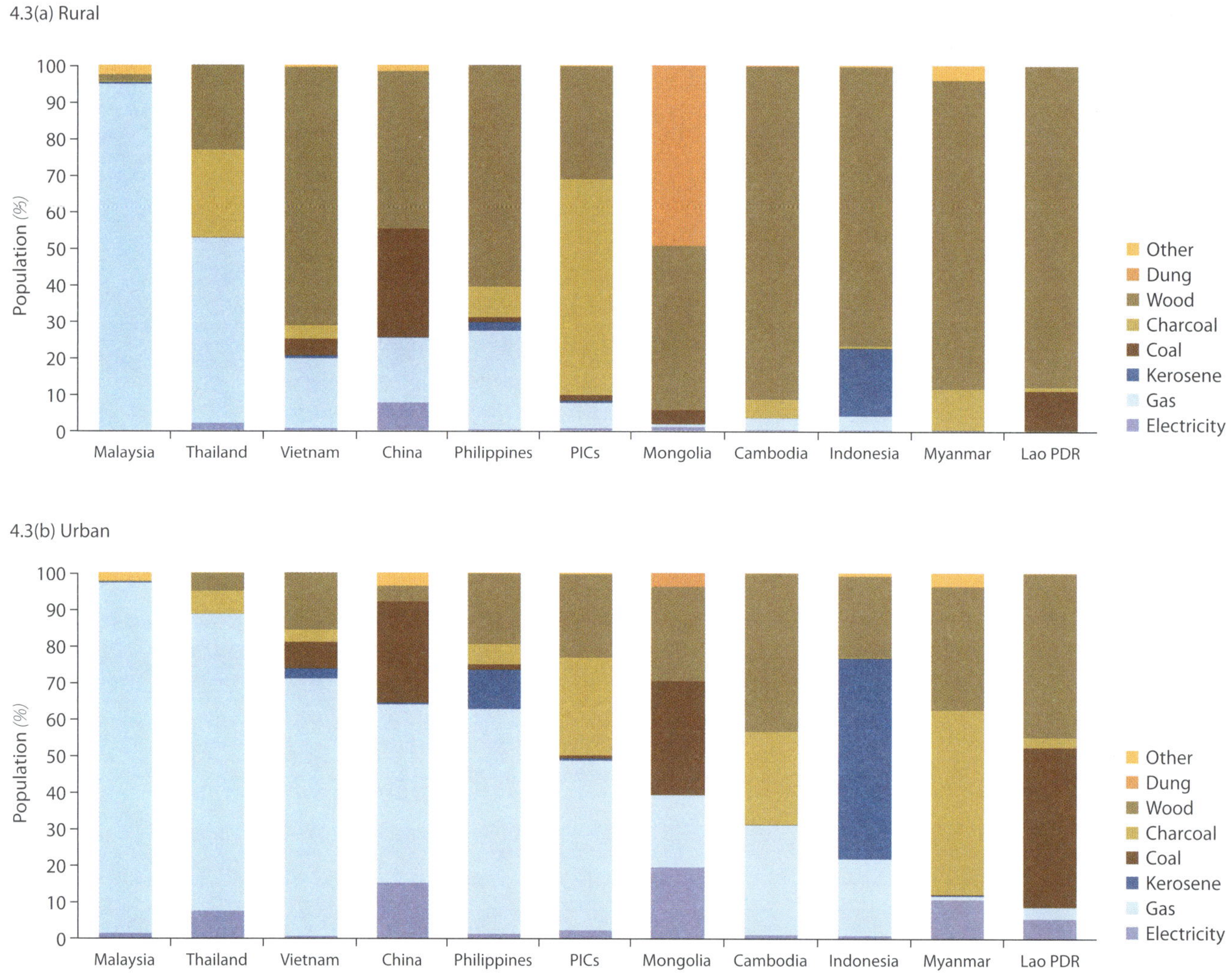

Source: WHO and UNDP 2009.
Note: The percentages are based on HH surveys conducted primarily 2006–08. Gas included biogas, piped gas, and LPG.

fact, there is evidence that people often switch to better types of cooking only in a partial and gradual manner. For instance, households may use LPG for boiling water and wood for cooking main meals before graduating completely to LPG (Masera and others 2000). Therefore, it is necessary to consider policies and programs that encourage faster adoption of modern fuels or stoves that offer more complete combustion of biomass fuels.

Urban and rural areas in EAP show clear income differences, making it important to tailor the promotion of modern cooking to each area. In urban areas, higher incomes and greater affordability result in greater use of modern fuels, whereas rural areas with lower incomes make greater use of transition or traditional fuels. During the coming decades, as urban incomes grow, the marketing of modern fuels by commercial companies will increase. Some countries such as Thailand already have reached almost universal levels of LPG use in urban areas. In low-income countries, if lower incomes and lower affordability persist even in urban areas, the use of transition

Table 4.3 Overview of Policies That Promote Clean and Efficient Cooking in Urban and Rural Areas

	Income and affordability	Fuel prices and subsidies	Access
Urban	***Overview:*** Higher incomes, good access, and greater affordability mean the higher use of modern fuels by high-income groups, and of transition or traditional fuels by lower income groups.	***Overview:*** All fuels are priced. Taxes and subsidies impact fuel adoption and use. Traditional fuel prices often follow the price of modern fuels so it is important not to tax modern fuels excessively.	***Overview:*** Modern fuels and transition fuels are available. Some countries have set fuel import limits. These generally should be avoided.
	Policies Improve affordability by providing microcredit or loans to lower upfront payments. Applicable grant programs also may be used to lower costs to consumers.	**Policies** *Modern and transition fuels:* Minimize taxes or subsidies. *Traditional fuels:* Prices are set by market but often adjust to price of modern or transition fuels. For this reason, modern or transition fuel taxes increase energy costs of the poor and should be avoided.	**Policies** *LPG:* Partial subsidies for stoves and cylinders. *Kerosene:* Partial subsidies for clean-burning stoves. *Solid fuels:* Technical assistance to develop and market stoves; partial subsidies for improved or advanced stoves.
Rural	***Overview:*** Lower incomes, poor access, and lower affordability mean greater use of transition and traditional fuels. More efficient improved or advanced stoves can lower fuel costs and improve affordability.	***Overview:*** Modern and transition fuels are priced above affordable levels for those with low income.	***Overview:*** Modern and transition fuels often are not distributed in rural areas, but traditional fuels are always available.
	Policies Improve affordability by providing microcredit or loans to lower upfront payments. Applicable grant programs also may be used to lower costs to consumers.	**Policies** *Modern and transition fuels:* Price at level similar to fuel prices in urban markets, perhaps through cross-subsidy mechanism that makes up for higher distribution costs. *Traditional fuels:* Let market set price; encourage local tree planting; and encourage local forest management to increase sustainable supply.	**Policies** *Modern and transition fuels:* Encourage greater distribution of modern and transition fuels and partial subsidies for stoves and/or cylinders. *Traditional and solid fuels:* Provide technical assistance for better stove development and partial subsidies for improved and advanced stoves. *Biogas:* Promote biogas systems to farmers with animals; include partial subsidies.

Sources: Derived from Barnes and others 2005; World Bank 1996; World Bank 2002.

Note: Modern fuels are electricity and any form of gas. Transition fuels are kerosene, coal, and charcoal. Traditional fuels are wood, straw, and dung.

and traditional fuels will continue to be high. In Cambodia, over 40 percent of the urban population still uses traditional fuels. For these reasons, it is appropriate to approach urban and rural energy in different ways when considering programs to promote clean cooking (table 4.3).

Prices: Modern Fuels and Interfuel Substitution

Fuel prices have a very significant impact on energy consumption patterns for cooking. In most rural areas, biomass energy is available from the local environment, and the cost is the labor time used to collect it. Therefore, it is not surprising that low-income HH in rural areas use biomass fuels for a large percentage of their cooking needs.

Petroleum-based fuels for cooking—mainly LPG—are preferred in higher income urban areas because these fuels are convenient and easy to use, even though they are more expensive than biomass fuels. Often no financial help is provided to the poor households because it is assumed that they eventually will catch up and switch to petroleum fuels. However, the cost differentials to do so are quite high. Thus, strong reasons exist for policymakers to seek attractive alternatives that use the same biomass fuels in more efficient and modern ways.

EAP countries have adopted a wide range of policies to tax or subsidize cooking fuels. These policies generally are related to the availability of the relevant natural resources in a country. If a country is required to import a fuel, the tendency is to tax it. If the fuel is produced within the country, subsidies are more likely. Until recently, kerosene subsidies in Indonesia traditionally were very high. As a result, the country stands out in the Region as using kerosene extensively for cooking in both rural and urban areas. The problem with keeping prices high is that the fuel inevitably is purchased and used in other sectors such as transport and industry (von Moltke and others 2003). In contrast, in China, coal is readily available and has a relatively low price. The consequence is that close to 30 percent of people in both urban and rural areas in China use coal as their main cooking fuel. Thus, it is better to have prices that are not significantly higher or lower than international benchmarks. However, in recent years, coal prices have steadily increased and may have encouraged a transition to improved cooking stoves.

Improving Affordability of Modern Cooking Fuels

The access to various cooking fuels often is related to import policies and the affordability of purchasing the stove and fuel. Until the early 1990s, to conserve foreign exchange and to use the readily available and relatively inexpensive coal in the country, China limited the import of LPG. This policy resulted in LPG being rationed informally so that even people who could afford the fuel could not purchase it.[30] Since the early 1990s, as the country opened its LPG market to competition from international investors, China's LPG consumption grew rapidly by an annual rate of 18 percent. At the end of the decade, China had become the world's third largest LPG consumer, following the United States and Japan. Presently, to conserve their foreign exchange for other goods and products, small countries in the Region including Cambodia and Lao PDR have very limited capacity to import cooking fuels. The nature of import policies is one reason that the majority of their populations remains dependent on solid fuels for cooking. The solution to such obstacles is to adopt appropriate cooking fuel import policies and perhaps provide subsidies or loans to the poor to pay the upfront costs of both modern and advanced solid fuel cookstoves.

Likewise, low-income households often cannot afford the upfront cost of an LPG stove even if they can afford to buy the fuel. Furthermore, LPG cylinders often contain a month's supply. Distribution of fuels in large container sizes can make these fuels unaffordable to poor people. Their response is to avoid adopting LPG. Instead, they purchase small amounts of kerosene several times a month. The solution to such obstacles is to adopt suitable cooking fuel import policies, and perhaps provide subsidies or loans to the

30. Tian 2002.

poor to pay the upfront costs of both modern and advanced solid fuel cookstoves.

To achieve both efficiency and equity, it is important for governments to rationalize taxes on modern fuels because the prices of modern fuels effectively set an upper limit for the prices of traditional fuels used by the urban poor (Barnes and others 2005). A tax on modern fuels also drives many middle-class HH to continue their reliance on wood beyond the point at which they otherwise would have switched to cleaner fuels. This tax puts additional pressure on the wood-based biomass resources around cities. Clearly, restrictions or bottlenecks on the distribution of transition fuels such as kerosene should be eliminated. Imports of petroleum products in limited quantities and subsequent subsidies and rationing should be avoided because the poor have difficulties in obtaining ration cards. Moreover, the limited supply means that the fuel has no cap effect on the price of fuelwood. Abandoning targeted subsidies and loosening fuel import restrictions may be needed to clear the bottlenecks or blocks to the adoption of transition fuels (Bacon and others 2010). Transition fuels are comparatively attractive and efficient, so it may be more productive to provide credits to low-income consumers to purchase efficient stoves than to subsidize the fuels purchased by the consumers.

Thailand has one of EAP's most successful programs to promote LPG use and has the highest level of LPG use in the Region. Thailand does have the advantage of being an LPG producer so is able to price the fuel below world market levels. However, the main reason for the success of the LPG program was that the government established an oil price stabilization fund for domestic petroleum prices. This fund was used to provide financing for storage facilities intended to keep the price of LPG in rural areas at levels similar to the price in Bangkok (box 4.2).

To summarize, with the exception of electricity, which is not used extensively for cooking, it will be necessary to work with oil and gas companies to promote LPG or kerosene for cooking. The established best practices for providing access to LPG for the people in the Region and elsewhere are:

- Possibly provide grants for stoves but not for fuel
- Distribute LPG in small bottles and make them widely available at a prorated price similar to LPG in larger bottles
- Devise financial and regulatory incentives to distribution companies to expand LPG markets
- Apply tax policies that do not make LPG or kerosene unaffordable or raise the prices of traditional alternatives such as fuelwood.

Advances in Cookstoves and Biogas Systems

Improved and Advanced Stoves

For rural areas, a major emphasis on marketing and promoting new efficient solid fuel stoves and biogas energy systems is needed. To date, there have been some very successful improved-stoves programs in the Region; in China, people have adopted over 100 million improved cookstoves (Sinton and others 2004; Smith and others 1993). However, even the best of these stoves still have a significant way to go to match the combustion efficiency of modern fuels. Today, manufacturers in China and some other EAP countries have begun producing newer high-efficiency and high-combustion stoves (figure 4.4). For rural areas, it also is possible to promote biogas energy systems for farmers who have animals. Biogas systems can provide clean gas for cooking and have been disseminated with some success in China and Vietnam.

Existing regional cookstove programs in China and Cambodia. While several EAP countries have conducted fairly successful improved stove programs, China's National Improved Stove Program (NISP) stands out as a remarkable success. It was implemented during the 1980s and 1990s by the county rural energy agencies, supported by the Ministry of Agriculture and other entities. The two focal points of this program were energy efficiency and removing indoor smoke through chimneys.

Box 4.2 Promoting LPG Use: Thailand's Successful Approach

The use of liquefied petroleum gas (LPG) for cooking in Thailand began in the mid-1970s. At that time, large petroleum companies such as Shell, Esso, and Caltex began selling LPG for households in Bangkok. LPG sold by these companies was largely a byproduct from their oil refineries in the country and a small amount from imports. However, the use of LPG initially was limited to a small group of higher-income HH in Bangkok. The reason was that LPG distribution was limited to Bangkok, and the price of LPG was very high compared with the prices of the widely available firewood and charcoal. In the early 1980s, the use of LPG gradually spread to upper-middle- and middle-income HH in Bangkok. The government began a series of efforts to promote the use of LPG for cooking by all HH in Bangkok and throughout the country. The price of LPG was set at the cost of production, which included regulated profit margins, rather than at world market prices.

In 1986, to promote the use of LPG among HH who lived in the provincial cities and rural areas, a government decree set a uniform wholesale pricing policy for LPG. The decree stated that wholesale prices at the five large regional storage facilities serving consumers outside the Greater Bangkok Metropolitan Area would be the same as the wholesale price in the Bangkok Metro Area. Using the fuel levy from Thailand's Oil Stabilization Fund, the government subsidizes the costs to transport LPG from the three main LPG storage facilities to regional storage facilities, which also serve as the distribution centers for those regions.

In addition, as part of the effort to promote the entry of new LPG distributors into the country, especially outside the Bangkok Metro Area, the government instructed the state-owned oil company, which owns storage facilities. The state-owned company allows other LPG suppliers/distributors and traders to use the company-owned storage facility free of charge. As a result, the number of LPG suppliers/distributors in the country has increased from 3 to 6.

Although the *amount* of LPG used for cooking has increased by an annual average of approximately 10 percent for the past 25 years, the *share* of LPG used for cooking declined from the peak of 78 percent in 1989 to approximately 50 percent in 2009. This decline was due to two compounding factors: the increasing use of LPG (a) for automobiles and (b) as the feedstock for the petrochemical industry for the past 10 years, caused in part by the low price of LPG in Thailand.

Source: Tuntivate 2010.

Over 100 million NISP improved stoves are still in use in the China. They have resulted in the largest improvement in energy efficiency among EAP countries and perhaps all regions (Smith and Deng 2010). China's experience supplies ample evidence that the development of a program for better stoves can succeed.

The two drawbacks of the NISP stoves are that their combustion efficiency is low and they cannot achieve a very clean burning. Consequently, although smoke was moved outside of the houses through the chimneys, thus resolving some of the indoor air pollution problems, the chimneys did not alleviate the general build-up of pollution levels in the communities.

The National Improved Stove Program ended in the late 1990s. With the exception of grant-funded programs in very poor areas, Chinese government support for commercializing stoves has ceased. Some manufacturers still are producing and selling the legacy stoves. However, more recently, the private sector began selling highly efficient and cleaner burning stove models (box 4.3). In fact, this development can be characterized as the emergence of a new generation of more advanced stoves. They are quite durable and have low emissions and high efficiency. Many of them have received significant consumer testing before being introduced to the public. Presently, advanced biomass stoves are being produced in limited quantities and are

Box 4.3 New Efficient and Cleaner Burning Stoves for China: Scope for Renewed Efforts

In China, with the extensive use of coal for heating in many regions, the market for commercial coal heating stoves is very strong. Each year, China produces approximately 2 million efficient coal stoves. Nevertheless, much larger numbers of such improved coal stoves must be produced soon to serve the nearly 50 million coal-using households in China.

In comparison to China's promotion of improved coal stoves, its promotion of more efficient and cleaner burning biomass stoves has lagged far behind. Most active government support for biomass stove programs has been curtailed. Moreover, the role of the county rural energy units, which actively supported the previous programs, has been assumed by the private sector. Many biomass-using households rely on the cooking technologies developed decades ago and still being disseminated by the private sector.

However, the development and production of new biomass stoves has not kept pace with the dimensions of the challenges to promote clean cooking. Approximately only 180,000 low-emission, high-efficiency biomass stoves are being produced and sold every year in China. Although the production of such stoves has been rising, it will take many years to reach the nearly 130 million households who use biomass as their main cooking fuel. China clearly needs to build on its earlier successful stove programs and undertake more active interventions. Options include developing an innovative new generation of cleaner burning biomass stoves, developing better marketing techniques to promote them, and encouraging the private sector to both market and sell a new generation of more advanced biomass stoves.

Source: Jia 2011.

not sold or promoted widely. The impacts that these stoves could have on both improving health and reducing drudgery could be quite significant for China. The possibility exists that if these new stoves were produced on a larger scale, their costs could be lowered significantly, enabling them to be sold widely in EAP. Already, some of these new advanced biomass stoves are being manufactured in China and exported to many parts of the world.

New possibilities exist to promote improved stoves on a wide scale in the Region. For instance, in Cambodia, GERES, an NGO, is working to convert traditional stove makers into improved stove makers and to turn traditional stove customers into consumers of improved charcoal stoves.[31] The GERES program also is one of the first improved stove programs to participate in the international carbon market. An important achievement of the GERES program has been to set up a viable supply chain for the manufacture and delivery of the New Lao Stove (NLS), while ensuring product quality and service (box 4.4).

While, in the past, there have been problems in implementing improved stove programs in EAP and around the world, there also have been successful instances from which to draw

Figure 4.4 Testing New Generation of Stoves in China

Photograph: Ashden Awards.

31. Groupe Energies Renouvelables, Environnement et Solidarités.

Box 4.4 Setting up a Supply Chain for New Lao Charcoal Stoves in Cambodia

Since 1994, GERES (Groupe Energies Renouvelables, Environnement et Solidarités), an NGO, has been working in Cambodia to develop energy-efficient solutions designed to conserve the environment and improve the living conditions of the Cambodian population. In 1999 GERES stimulated the introduction of the New Lao Stove (NLS), supported by trainers from Thailand, where it was being marketed under the name, "Thai Bucket." The technical design of the NLS is an updraft combustion stove with a grate. After training a group of existing cookstove producers, Lao PDR began the initial comparative tests against the competing traditional model known as the Traditional Lao Stove. The New Lao Stove is a charcoal stove, but similar stoves are manufactured by international firms specializing in wood stove development and sales.

The more innovative aspect of the GERES effort is the successful institutional model for selling the stoves, rather than the design of the stove itself. One challenge that emerged during project implementation was the large number of decentralized production units. Having approximately 31 scattered production centers made it difficult to control the quality of the stove. GERES addressed this challenge by consolidating the scattered production centers to five centralized facilities. Next, the NGO set up a local supply chain, selected a trial area, and trained producers to produce stoves first for this area, and then for the country. The entire supply chain has the extensive participation of women, who are managing retail shops and promoting stoves.

Box figure 4.4.1 Cooking with New Lao Stove

Photograph: www.geres-cambodia.org/

From 2003 to 2010, the sales of the NLS in Cambodia totaled 1 million units, well ahead of projections. According to the manufacturer, compared with traditional stoves, the New Lao Stove can save a considerable quantity of charcoal. Due to its proven ability to reduce CO_2 emissions, in 2006 GERES Cambodia was the first project developer in the world to put forward an improved cookstove project to trade on the carbon market. The price of the stove is US$2–$4.

Sources: www.geres.eu; World Bank 2011.

lessons. Furthermore, the outlook for improved stoves is encouraging as new varieties of stoves are developed and new alliances are formed to break through the barriers to wide dissemination. Expanding such efforts would greatly benefit the Region.

Biogas Systems, Gasification, and LPG: Encourage Commercialization and Marketing

The use of biogas systems for cooking is based on methane gas that can be used in a manner similar to LPG. The biogas system is a small niche technology that can transform biomass into clean-burning gas for modern cooking. The technology is well proven, although the use of biogas is limited to farmers who have two or more farm animals. In addition to new varieties of cookstoves, smaller niche cooking technologies such as biogas systems can play significant roles in improving cooking practices (REN21 2010). The introduction of biogas for cooking has been a slow but steady process in developing countries, in part because the manure feedstock limits the market for household biogas systems to animal owners.

However, after roughly 25 years of design experimentation, the technology is entering a new phase. China now has some 25 million biogas systems, with an estimated 3 million added

Box 4.5 Biogas Energy and Carbon Financing in the World Bank's Hubei Eco-Farming Project, China

The Government of China sees biogas use as a means to improve the lives of rural households and address global and local environmental issues. For these reasons, in 2001 the country launched a large National Rural Biogas Program. To date, investments total more than RMB 3 billion (approximately US$375 million). This program has resulted in 7.2 million rural households in China cooking with biogas.

Recently, a World-Bank-funded Eco-Farming Project was initiated to complement the government's efforts and support further expansion of the rural biogas program. The project is located primarily in the region of Hubei, a rural mountainous district in China whose main economic activity is agriculture. Households in this region are being given the opportunity to install biogas digesters that use animal waste to generate gas for use in domestic heating, lighting, and cooking. Under the Carbon Development Mechanism (CDM) component of the project, more than 33,000 households—or approximately 165,000 people—are benefitting from the installation of domestic digesters, which are displacing more carbon-intensive traditional domestic fuels such as firewood, coal, coke, and crop residues.

The Eco-Farming Project also assists farmer households to integrate biogas in their agricultural production systems to improve their environmental impact, quality, and efficiency; and to improve the households' living conditions. For the sustainable operation and maintenance (O&M) of the biogas systems, the project will (1) strengthen rural energy and agricultural extension services as well as local extension and training facilities, and (2) provide equipment and materials for the service systems; training for counties, townships, farmer technicians, and farmers; and support to farmer biogas organizations.

In addition, the project illustrates how the CDM mechanism can be leveraged effectively through biogas applications to:

- Reduce the time spent by women collecting firewood or travelling to purchase fuel
- Significantly reduce the amount of household income spent on fuel
- Improve indoor air quality in homes due to the smoke-free combustion of biogas compared to traditional fuels
- Enhance sanitation services by improving swine manure handling techniques, as well as providing a connected latrine for households.

Sources: UNFCC 2010; World Bank 2008c.

during 2009 (REN21 2010) (box 4.5). Vietnam has more than 150,000 systems. Outside EAP, Nepal's Biogas Support Program combines the participation of the private sector, microfinance organizations, community groups, and NGOs. During the last decade, this program steadily has increased biogas systems: close to 200,000 systems have been implemented.

The use of LPG has been expanding rapidly in China, Thailand, Vietnam, and other countries in the Region. Continuing the incentives to use LPG is a wise policy. Large companies are not likely to be interested in promoting LPG in rural areas due to the costs of extending the supply network and the relatively low incomes and affordability on the part of rural consumers. Consequently, there is a strong need for the government to provide incentives for expanding the availability and use of LPG through appropriate policy interventions.

Two Scenarios for Modern Cooking Solutions

Due to increasing urbanization, by 2030 over 50 percent of the EAP population will reside in urban areas. This report has developed two scenarios for both urban and rural areas that predict fuel use for 2030 based on the changes in

Figure 4.5 Baseline (2010) and Projected (2030) Cooking Fuel Use in EAP

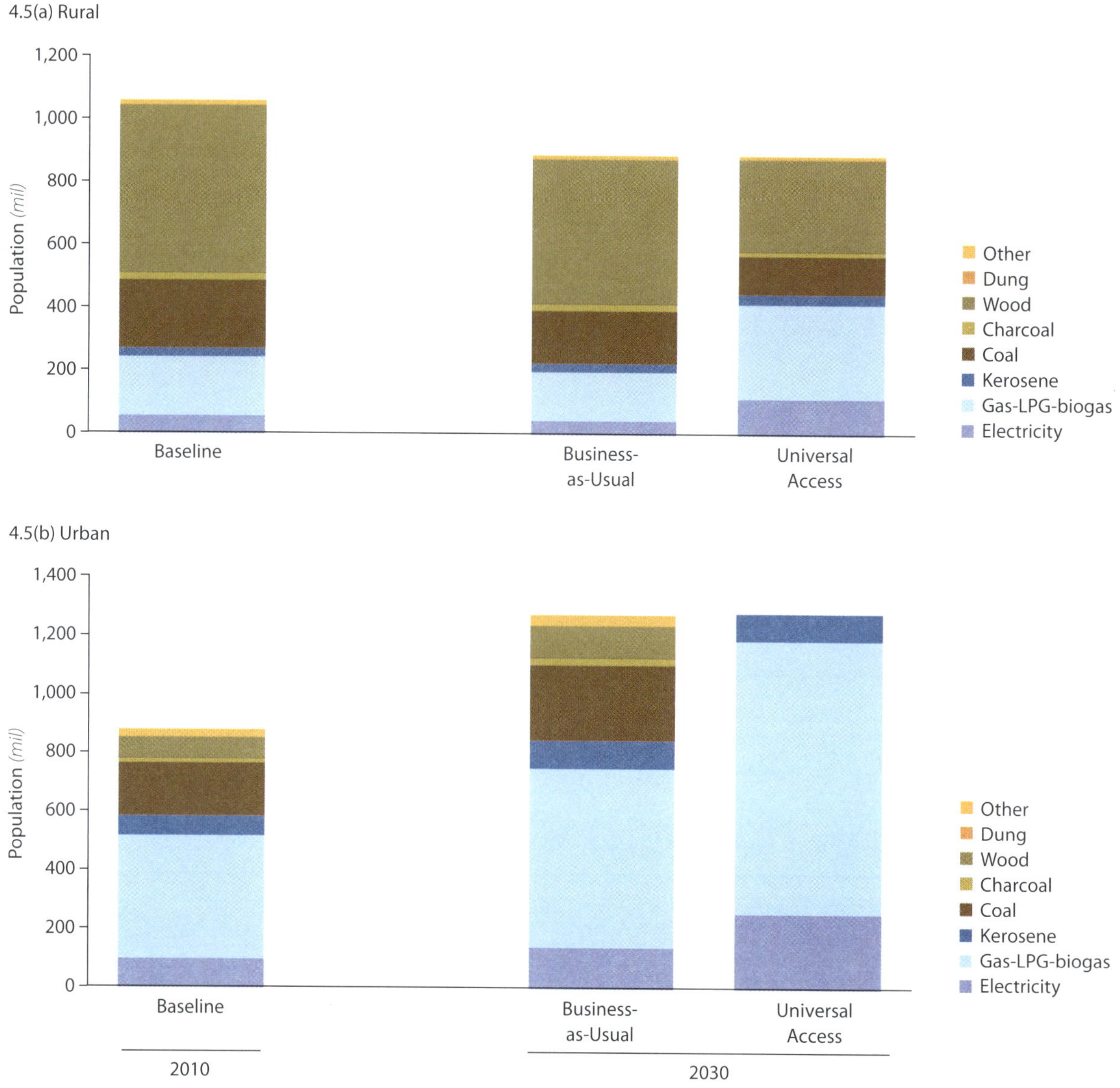

Sources: WHO and UNDP 2009; UN-DESA 2008; authors' calculations.

population and rural-urban population patterns. These scenarios closely resemble those for electricity but are more complicated due to the complex relationship between income growth and interfuel substitution. The first scenario is relatively conservative and involves interfuel substitution caused mainly by urbanization. The second scenario is for near-universal access to clean cooking fuels, and involves significant increases in the rural use of LPG, biogas energy, and better cookstoves. The following sections define these scenarios more precisely. (See also figure 4.5.)

Business-as-Usual Scenario

The Business-as-Usual scenario for clean cooking involves no growth in the percentage of EAP households using modern fuels (mainly LPG and electricity) in urban or rural areas. However, due to increasing urbanization, the absolute number of people using modern fuels will increase.

Under this scenario, by 2030 rural areas still will have 665 million people who use solid fuels for cooking. Of these, 460 million still will depend on wood straw or dung. In contrast, in urban areas, only 100 million people still will be using traditional fuels and 250 million using coal for cooking, for a total of 350 million people who will not be using modern fuels.

Universal Access Scenario

For the Universal Access scenario, this report assumes that by 2030 the entire population in EAP urban areas will be using modern fuels (electricity, LPG, or kerosene). However, rural populations will make only a partial switch to modern fuels due to lower incomes and the lack of availability of modern fuels. Overall, compared to the Business-as-Usual scenario, under Universal Access, by 2030 an estimated additional 435 million urban households will be using modern energy. Moreover, and importantly, the rural population using traditional fuels will have adopted clean and efficient stoves. For rural areas, compared to the Business-as-Usual scenario, Universal Access assumes approximately 20 percent growth in the use of modern fuels. Universal Access also assumes that 300 million rural HH still will be using traditional fuels, and over 130 million still will be cooking with charcoal or coal by 2030.

According to these scenarios, two quite different types of challenges exist. The urban areas will gain population so must develop the requisite energy infrastructure, mainly through the types of energy that are provided through large commercial networks. This energy includes electricity, LPG, and kerosene. In contrast, in rural areas, some growth of cooking with LPG and kerosene is expected. However, it also will be necessary to address the problems arising from the use of solid fuels as the main cooking sources (table 1.4). This challenge will be to address the issue of clean cooking in both urban and rural areas with related but somewhat different strategies.

Urban challenges. The main challenges in urban areas will be to develop the commercial fuel infrastructure to serve approximately 230 million new customers under the Business-as-Usual scenario and 430 million new customers under the Universal Access scenario by 2030. These challenges will require significant levels of increased electricity generation as well as investments in developing importing and processing facilities for gas and LPG. These investments could generate new market opportunities for commercial development of these fuels. However, expanding decentralized market infrastructure is much more difficult than serving large customers. For some of the smaller countries, foreign exchange may be a constraint. Finally, to achieve their energy access goals, many countries in the Region may have to consider reforming their pricing and related policies regarding modern fuels.

Rural challenges. Difficulties are even more pronounced in reaching out to rural customers in remote areas. The comparatively low population densities of rural areas means that commercial fuel companies often do not wish to serve in such areas due to high costs, low-income consumers, and the relatively low number of new consumers. However, even more challenging will be the development of BoP models to market and introduce large numbers of improved solid fuel stoves in rural areas. To scale up, governments must address six fundamental issues (World Bank 2011a):

1. Few countries have national institutions/agencies to promote cookstove programs.
2. Stove models suitable for BoP populations have been developed and continue to evolve, but these models have not yet been disseminated on a large scale.
3. Existing loan funds administered by financial groups generally are not used to finance clean and efficient biomass stoves.
4. No well-accepted standards, or even quasi-standards, exist to qualify the stoves as safe, durable, efficient, and clean burning.
5. Support for the technical development of clean and efficient stoves is limited.
6. Most petroleum fuel companies in the Region lack incentives to provide access to such fuels as LPG in rural areas.

If they are to achieve universal access to clean and efficient cooking in the next 2 decades, EAP countries must address these 2 challenges of achieving a high level of commercial fuel marketing and scaling up the use of better solid fuel stoves.

Financing Requirements for Clean Cooking

The investment projections under the 2 scenarios are based on the projected population growth from 2010 to 2030. The Business-as-Usual scenario assumes that interfuel substitution will occur passively, primarily from urbanization. In contrast, the Universal Access scenario assumes that governments will make significant shifts in policies to actively encourage better cooking practices. The financing for these scenarios will come mainly from the households who purchase the cooking fuels. However, some government interventions may be required to promote this transition to more modern cooking methods. For both the urban and the rural scenarios, the costs of the cooking practices are assumed to be similar. Thus, the differences between urban and rural areas will be the different types of fuels and the different numbers of people according to the population projections.

Even though it is recognized that a stove will not last 20 years, the costs are based on the cost of only the first stove that a household buys because it is difficult to estimate the number of stoves that a household will need over 20 years. The assumed prices are US$100 for modern fuel stoves (such as LPG stoves) and US$50 for high-quality solid fuel stoves. The investments necessary for LPG do not include the associated new storage or distribution costs required to enable retailers to sell the fuel to the urban poor or those in rural areas. Thus, these investment figures are approximations that are intended primarily to indicate the minimum investments necessary to promote a clean cooking agenda. The modern fuel stove investments are intended only for the incremental stoves being purchased by new users who are part of rapid urbanization.

The total investment costs of the Business-as-Usual scenario are an estimated $16 billion (figure 4.6). In rural areas, the total would be approximately $6.6 billion, including $4.6 billion for wood stoves and $1.7 billion for coal stoves. In urban areas, the cost of the stoves would total $9.5 billion, including $1 billion for wood stoves, $2.5 billion for coal stoves, and $4.4 billion for modern fuel stoves. Under the

Figure 4.6 Investment Needs for Modern Cookstoves under Business-as-Usual and Universal Access Scenarios by Cooking Fuel

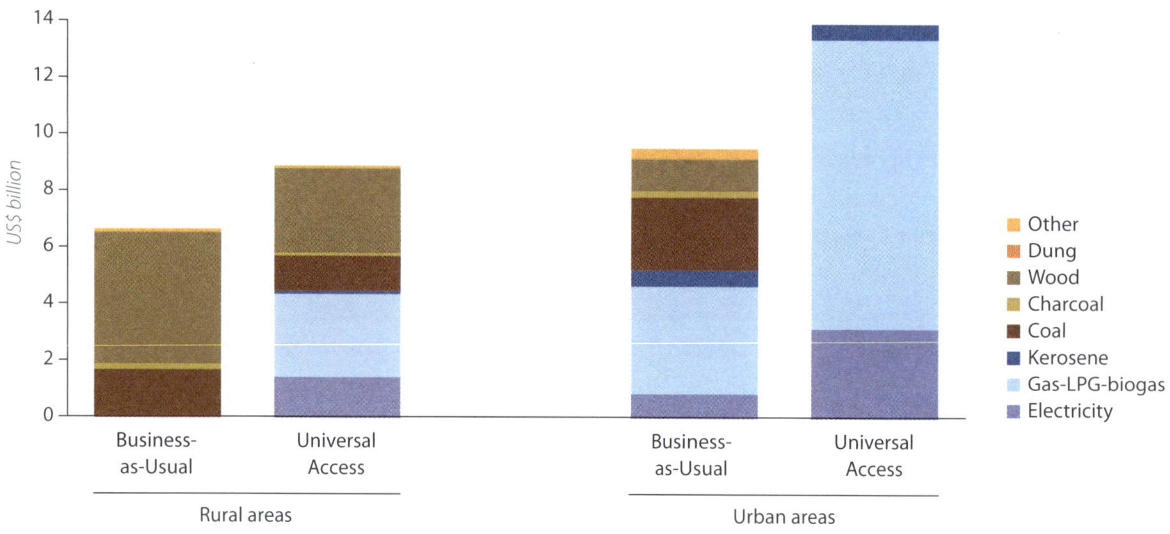

Sources: WHO and UNDP 2009; authors' calculations.

Universal Access scenario, the required investments rise from $16 billion to $22 billion. The additional investments (above the Business-as-Usual scenario) for urban areas by 2030 would be $4 billion; and for rural areas, $2 billion. Promoting the modern fuel stove would cost approximately $4 billion dollars.

Given the potential benefits, these are not excessively large investment amounts, Most of these expenses will be borne by household consumers. However, to move from the Business-as-Usual scenario to the Universal Access scenario will require both innovations in the way that the problem is perceived by policymakers and technical innovations in the way that household energy services are delivered to consumers. It is obvious that new investments are required to move from the Business-as-Usual to the Universal Access scenario for EAP households.

Implementation Requirements

Efforts to improve access to modern fuels require cooperation and coordination among multiple sectors: rural, forests, environmental, social, and financial. The deployment of improved cookstoves is a logistically challenging exercise that requires coordinating appropriate technology and designs, distribution channels, financing mechanisms, and service support. As part of overall Regional energy plans, the problems associated with the high number of people cooking with solid fuels in the Region need to be more actively addressed.

Similar to the electricity sector, for modern fuels, large government or private commercial enterprises already exist that deal with modern fuels such as kerosene or LPG. However, these enterprises generally lack sufficient incentives to extend their market reach to the poor in urban areas and to rural areas in which population densities and commercial markets are less desirable. Thus, countries that have lagged in expanding the use of modern fuels should put in place incentives to encourage firms to reach out to the people who still use solid fuels with traditional means of cooking.

To develop better solid fuel stoves, there would have to be a period of technical assistance for market and technical stove development. It would be followed by a period in which government would encourage improved stove adoption through partial subsidies or microcredit loans for the stoves themselves. Assuming a commercial approach to the promotion of clean cooking, governments generally need to focus on the ways to promote market development for interventions that are clean, efficient, safe, and reliable. It is likely that such promotional efforts would require high levels of TA and specialized institutions to deal with the problem. In addition, for such small interventions, generally the private sector, NGOs, and microfinance organizations would have to market and promote the stoves, so a period of support would be necessary to strengthen the capacity of these organizations to undertake such activities.

Cooking with biogas, Vietnam.

Photo credit: Ashden Awards.

5

Modern Cooking Solutions: The Way Forward

The efforts to provide universal access to modern cooking solutions will include at least three focal points: expansion of LPG, promotion of clean and efficient stoves, and development of biogas energy systems, with the last limited to HH with the requisite number of farm animals. To date, the promotion of LPG probably has been the major approach to advance clean cooking in the Region. The wider dissemination of LPG is dependent primarily on governments' pricing and import policies (chapter 4).

This chapter focuses on promoting clean and efficient cookstoves and biogas systems among the rural and urban poor, who are unlikely to be able to afford LPG in the near term. Since biogas energy systems can serve only a limited number of people, the key to delivering clean cooking systems is the development and dissemination of improved cookstoves that use traditional biomass and coal as fuel.

Even though efforts to promote clean cookstoves have been ongoing for many decades, programs in EAP and worldwide have been limited. China and India have undertaken large-scale programs, but most other countries have not advanced significantly beyond pilots. However, even where the pilot programs have succeeded, the commitment to clean cooking often has been limited and fragmented.

In recent years, new approaches to promoting modern cooking solutions have emerged, along with a growing consensus on the urgency of this effort. Furthermore, there have been some successful programs in EAP from which lessons can be derived for future programs. For instance, China's successful NISP has disseminated over 100 million improved stoves (Sinton and others 2004). This success was achieved through learning by doing and adjusting early unsuccessful methodologies. In Thailand, before LPG became the predominant fuel, the efficient bucket stove became almost universally adopted by people in urban areas (Kammen 1995).

The emerging new approaches and lessons from past programs provide the foundations on which to design future programs in the Region. It is essential to begin now because most EAP countries face a long road ahead to adopt and promote the best policies and strategies for achieving universal access to modern cooking solutions.

What Is Different from the Past?

It is widely accepted that many past cookstove and small-scale renewable energy programs had several shortcomings in their technical designs and distribution strategies. The three major problems were that (1) most products were not well designed; (2) there was little quality control

Box 5.1 New Advanced Biomass Cookstoves Initiative in India

In 2009 India announced a large initiative for advanced biomass cookstoves. This initiative was intended to reach the millions or even hundreds of millions of people who use biomass energy and enable them to meet their daily cooking needs more healthfully. Under development, the program will focus on five key areas:

- Technical issues related to testing and standards, including R&D
- Cookstove delivery procedures
- Potential programs to process and supply fuel
- An innovation contest for next-generation cookstoves
- Exploring what can be accomplished with community cookstoves.

Expected in 2011, a new protocol will be adopted to test and qualify cookstoves based on their energy efficiency and stove emissions. Testing facilities will be set up prior to qualifying the cookstoves. It is expected that new manufacturers will be involved. Subsidy design may include carbon credits.

Source: IIT and ERI 2010.

during installation; and (3) some cookstoves had short working lives because they were built with limited expertise entirely of local materials. Ironically, these problems arose from well-intentioned efforts to use local entrepreneurs to manufacture them, to keep costs low, and to ensure that the stoves were appropriate for local cooking practices. Many of the stoves did work well when new, but their performance quickly degraded, and the stoves began to crack and break down. As a result, most of these past programs, with the notable exception of those in China and India, remained at a small scale and often were evaluated as problematic and unsuccessful.

Today, the situation is quite different. In the last decade, there has been an emergence of new technologies, new ways of producing stoves, better quality control, innovative financing schemes, and new coalitions to support the scaling up of new programs (World Bank 2011a). Some new stoves are being manufactured in factories and workshops and delivered ready to use. Other manufacturers are producing essential stove components such as fireboxes and chimneys in the form of kits that can be assembled onsite.

In 2009 India began a renewed effort to develop a successful stove program to meet the needs of the hundreds of millions of people who use biomass energy for cooking. That year, a new initiative was started with the goal of getting most people who use solid fuels to begin to use a new stove in the not too distant future (box 5.1).

India's cookstove initiative highlights the necessity to develop innovative stoves that are subject to standards and testing methods that qualify them for the new effort. The concept is that stoves should be certified to be safe, reliable, efficient, and clean burning. The initiative also may have scope for different levels or tiers of certification or stove labeling. Certainly, methods to finance the high costs of stoves are an issue that must be addressed along with the appropriate levels of subsidies for the stove itself. The research on the health benefits of reducing emissions is still an area that policymakers should emphasize. Today, all stakeholders have the opportunity to join together to tackle energy and indoor pollution problems that intensely affect the world's poorest populations.

Policy and Technical Support for New Approaches

Implementation of new approaches for better stoves requires several different types of support. To achieve the goals for cleaner cooking in the Region requires a new coalition of government, donor, private sector, and nongovernmental entities. Governments can play a supporting role as catalysts to promote the development of innovations, as opposed to providing heavy subsidies for a particular type of stove. This support will include providing incentives to both private sector and nongovernmental organizations to develop better products through the interaction among the designers, manufacturers, and users of stoves to produce more diverse stove designs that appeal to consumers. To enable this interaction, a technical infrastructure must be put in place to test the claims of new product manufacturers and to ensure that products deliver on performance promises and are safe and reliable. In addition, once satisfactory products become available, governments and other entities will have to launch consumer information campaigns to inform the public of the health and efficiency advantages of the new products.

The requirements to promote small retail products are fairly similar whether the latter are better stoves or more efficient lighting appliances. These requirements include encouraging innovative product designs that are significant improvements over past methods of cooking, lighting, or heating. In all likelihood, the new designs will emerge from small private companies or NGOs. Product development often is very expensive and could be supported through competitive innovation grants or other financial incentives provided by governments or other entities to assist with such expenses. The general requirements are to develop a system of technical support to ensure that (1) stoves perform as expected, and (2) financial and possible subsidy arrangements are put in place to support larger scale program development along with institutional support for implementation (World Bank 2010c).

Technical Support, Performance, and Information Exchange

Technical support certainly is needed to certify stoves to ensure that they perform satisfactorily. Certified stoves could be considered eligible for government financing or partial grants to support dissemination. Certification would require accepted country or international standards for improved stoves. Today, no such generally accepted standards exist worldwide, but some EAP countries have developed their own standards. Therefore, a Regional review and recommendations concerning acceptable standards for advanced biomass stoves are necessary to complement any existing country-level work in the Region (MacCarty and others 2007; MacCarty and others 2010). This Regional review would have to be coordinated by an entity that is directly involved in, or at least knowledgeable about, new work at the international level.

At present, over 100 different new types of stoves are estimated to be available in developing countries. These products vary substantially in their sophistication, durability, and reliability. As a consequence, suitable codes or standards need to be established to ensure that products meet the needs of consumers and perform satisfactorily. Thus, there is a need to test new products, while making sure that standards are not so stringent that they discourage the products' commercial development.

In the initial stages, technical support will be necessary to encourage better and more diverse products. For such small interventions as clean and efficient stoves, it is best to encourage a diversity of products to enable consumers to choose what best fits their needs. This diversity will require encouraging interaction among stove users, builders, and designers. In addition, governments and other entities should support significant monitoring and evaluation (M&E) of stove performance to ensure that lack of stove durability, a significant issue in past programs, does not remain a problem. Technical issues that should be addressed include stove durability, safety, and performance. To encourage product innovation, it would be a good idea to set up a technical exchange for information sharing on the issue of clean cooking. This exchange could be combined with the commercialization of lighting and other small household-energy related products.

Financial Framework to Encourage Production, Development, and Sales

The financing requirements for initiating any new agenda for clean cooking are fairly similar to programs that market new products such as solar home systems or microlighting systems. Many past cooking programs have suffered due to their pilot nature or short-term focus. To give them a greater chance to succeed, larger programs will require longer term government and donor commitment.

Subsidies will be necessary to promote the development of new stoves. Generally, the most effective stove programs have financed stove development costs, with minimal or no subsidies for the stove itself. In many countries, high subsidies for stoves themselves have proved problematic (World Bank 2011a). In the early stages of the NISP in China, the government provided significant subsidies for stoves, and stoves were promoted through county rural energy offices. After initial difficulties, China altered its approach to include support to develop and promote the stoves. Their distribution was accomplished without subsidies. This approach resulted in the establishment of factories to produce the stove components, which then were assembled by local entrepreneurs. The success of this program meant that China had no need to continue its past level of support for stove development.

There could be a role for vouchers that provide increasing subsidies based on the smoke removal or energy efficiency of stove products. The necessary balance between loans and grants to promote better stoves in developing countries will need to be evaluated in each situation, taking both cost and affordability into consideration.

The financing and subsidies involved in cookstove promotion programs are complex because the stoves generally are sold by private companies. Programs such as Lighting Africa have addressed such issues for electricity access (Lighting Africa 2011),[32] and lessons can be derived from their experience.

Furthermore, efforts should be made to explore the use of climate change funds so that the financing of advanced stoves is broadly mainstreamed. The Global Environment Facility (GEF) has listed clean cookstoves as eligible for grants. Methodologies already are available for voluntary carbon markets and are beginning to be used, although not yet widely (box 5.2).

Specialized funds also have been used successfully to promote better stoves and other technologies. Rural energy funds, which have been used in many countries to support rural electrification, can be used to promote improved cookstoves. Such funds are, by their very nature, open to new types of projects and use both NGOs and the private sector to implement projects. Private sector entities and NGOs could be employed to coordinate efforts to deal with these very important issues. Generally, specialized funds are flexible and also could be used by, or coordinated with, other sectoral programs such as clean water, rural development, and local forest management.

In addition to rural energy funds, many countries have community block grants or social funds that could be used to promote clean cooking and other energy technologies. One very successful program in Guatemala provided communities with many development options, one of which was to use the funds for improved stoves. These stoves had large iron or steel cooking surfaces that were convenient for cooking with pots, making flat breads, and roasting corn (Ahmed and others 2005).

Indonesia's Green PNPM (National Program for Community Empowerment) is an environmental pilot that provides block grants and technical assistance to communities (World Bank 2010b).[33] At present, the main grant products are forest and water resources management, environmental services, waste management, and renewable energy. With the right type of improved

32. Lighting Africa, a joint IFC-World Bank program, is helping to develop commercial off-grid lighting markets in Sub-Saharan Africa as part of the World Bank Group's wider efforts to increase access to energy.

33. PNPM Generasi is an innovative pilot program launched by the Government of Indonesia in July 2007. PNPM aims to accelerate achievement of three Millennium Development Goals: Universal basic education, Reduction in child mortality, and Improvement in maternal health.

Box 5.2 Potential for Financing Efficient Biomass Stove Projects through Climate and Carbon Funding Mechanisms

Several financing mechanisms designed to mitigate climate change can be leveraged to fund biomass energy projects including the development and deployment of efficient cookstoves.

Clean Development Mechanism (CDM). To date, the CDM has approved two methodologies for improved cookstoves and reduction of nonrenewable biomass: (a) AMS (Approved Methodology for Small-Scale) II.G and (b) "Methodology for Improved Cookstoves and Kitchen Regimes" (Gold Standard V.01). The AMS methodology is based primarily on carbon mitigation. The Gold Standard is more onerous. It uses a fossil-fuel baseline to compute the expected fossil-fuel savings from the local fossil alternative to determine the scale of financing. The Gold Standard methodology, used in voluntary carbon markets, also accounts for the inclusion of upstream emissions reductions, that is, emissions from charcoal production, as well as methane and nitrous-oxide emissions reductions.

With the ongoing simplification of the CDM process, it is becoming easier to develop smaller scale biomass energy projects such as with improved cookstoves. Additionally, CDM-certified credits traded on the emissions trading system (ETS) usually carry a premium over voluntary emissions reductions (VERs) credits. The CDM already has approved 26 household energy-efficiency projects and has registered 6 more.

Climate Investment Funds (CIF). Channeled through the multilateral development banks (MDBs), CIFs comprise the Clean Technology Fund (CTF) and the Strategic Climate Fund (SCF). SCF, in turn, covers two programs that may be used for scaling up forests and other biomass energy projects. These two programs are the Forest Investment Program (FIP) and the Scaling Up Renewable Energy Program (SREP). FIP supports developing countries' efforts for REDD (Reducing Emissions from Deforestation and Forest Degradation).

The World Bank Group. The World Bank (IBRD) houses three funds with particular relevance to biomass energy: (a) the BioCarbon Fund (BioCF), (b) the Community Development Carbon Fund (CDCF), and (c) the Forest Carbon Partnership Facility (FCPF). The first two funds focus on land-use-based credits and rural-community-based projects, respectively. FCPF is designed to support the Bank's efforts to address REDD. The International Finance Corporation (IFC) also has helped to develop a large number of finance products for businesses and consumers that could be used to support biomass energy. IFC recently also agreed to provide financing to support improved cookstoves.

Global Environment Facility (GEF). The GEF has three grant mechanisms to promote better biomass stoves and improve the sustainability of household biomass use: (a) the Earth Fund (and other private sector development funds), (b) the Sustainable Forest Management program, and (c) the Small Grants Program.

Sources: World Bank 2011a; UNEP Risoe Center 2010; www.ifc.org; GIZ 2011.

stoves, this program could be expanded to include the financing of an efficient and durable stove. In addition, a wide variety of global funding sources, especially climate and carbon funding mechanisms, could be harnessed to support clean cooking (box 5.2).

To summarize, cookstove finance can be broken down into components that address different needs. These divisions are important because methods and strategies need to be tailored to address the varied goals of cookstove programs. The program funding types and objectives include:

- Rapid-deployment financing to successful cookstove project developers to scale up existing projects

- Market development activities to create robust markets for advanced cookstoves in priority countries and regions
- Pilot programs to assess the technical performance and market viability of advanced stoves that deliver the best local (health) and global (climate) benefits
- Support to cookstove entrepreneurs and manufacturers in the form of funding and TA to foster the quality and quantity of advanced cookstoves in the market and to reduce costs
- Policy support to country governments to create enabling policy environments and to direct public sector resources to the problems that advanced cookstoves can address
- Humanitarian assistance in disaster and conflict zones to provide cookstoves to distressed populations, such as residents of refugee camps.

These various needs could be financed by a diverse set of organizations including multilateral and bilateral donors, international carbon funds, country governments, and private foundations.

Possible Global and Institutional Support to Clean Cooking

The promotion of modern cooking solutions has links with the energy, environment, health, and social sectors. Consequently, responsibility for this work does not fall squarely on any one sector. This multisectoral aspect is similar to the challenge in the water and sanitation sector, in which handwashing, smaller latrines, and toilets cut across the water, health, and environmental sectors. This multisectoral characteristic has been addressed by developing the multisectoral, grant-based Water and Sanitation Program (WSP), which is responsible for promoting projects within countries and creating international consensus for action (box 5.3). A similar arrangement can be considered for dealing with the multisectoral nature of improved cookstoves. In the past, the Energy Strategy Management Assistance Program (ESMAP) has supported a considerable amount of work on biomass energy and could play a significant role in assisting the EAP Region's cookstoves programs.

Several different types of institutions are needed to scale up clean cooking in EAP. One requirement is an institution that supports policy studies and project preparation. The World Bank's Asia Sustainable and Alternative Energy Program (ASTAE) is well suited for this type of work. As an example, much of the project development work for small market energy interventions in both East and South Asia was the result of work financed by ASTAE grants. ASTAE

Box 5.3 Water and Sanitation Program: Potential Model to Promote Modern Cooking Solutions

The challenge for modern cooking solutions is to promote and replicate successful approaches, continue targeted learning efforts, and support reforms that ensure the adoption of sustainable policies and investments across the relevant sectors: energy, environment, health, and social. The Water and Sanitation Program (WSP) presents a useful model. WSP is a multidonor partnership administered by the World Bank to support poor people in obtaining affordable, safe, and sustainable access to water and sanitation services. WSP works directly with client governments at the local and national levels in 25 countries through regional offices in Africa, East and South Asia, Latin America and the Caribbean; and in Washington, DC. Over the last three decades, WSP has led or supported many of the advances made within the water and sanitation sector. It has been able to share best practices across Regions and place a strong focus on capacity building by forming partnerships with academia, civil society organizations (CSOs), donors, governments, media, the private sector, and others. WSP's work helps to effect the regulatory and structural changes needed for broad water and sanitation sector reform.

Source: www.wsp.org

could provide financing for project innovations and policy work on how to scale up clean cooking in EAP. Moreover, because ASTAE also covers the South Asia Region, EAP could provide useful lessons for projects developed for all of Asia. In fact, ASTAE already supports a limited number of project interventions to promote better stoves and clean cooking. In short, ASTAE could be one of the focal points for intervention within EAP.

The promotion of better cookstoves in EAP is a large-scale intervention that will require a *new or existing institution* to take on the challenge of moving the Region's 1 billion people to more modern and clean ways of cooking. Any new institution would have to be supported by the country governments. Its primary role would be the development and exchange of information to promote clean cooking solutions. All of these issues need to be addressed in EAP to support the expansion of efforts to promote clean cooking.

In 2010 the United Nations Foundation launched the Global Alliance for Clean Cookstoves to advocate for the implementation of a variety of programs for better biomass stoves in developing countries (UN Foundation 2010; World Bank 2011a). The issues featured in the Global Alliance are similar to those that are facing EAP countries. The types of activities encompassed by the Global Alliance to scale up and deploy clean cooking solutions include:

- Development of standards and testing methods
- Ways to encourage adoption of advanced stoves
- Ways to spread out the upfront cost of stoves through securing appropriate financing
- Filling in some of the major existing research gaps that will be necessary to promote the program
- Development of effective awareness-raising methods
- Promotion of market-based solutions that include the private sector, and nongovernmental and microfinance organizations.

EAP countries have a variety of institutions that already implement cookstove programs. China's Ministry of Agriculture has been the focal point for improved stoves. Mongolia's new energy institution (Energy Authority) could be adapted to cover clean cooking as part of its work. However, in most EAP countries, the programs for clean cooking are small and sometimes ineffective. It is essential that individual countries commit to promote improved cookstoves, and nominate/develop capable institutions that will take overall responsibility for the successful implementation of improved cookstoves programs.

Fortunately, many effective NGOs already are promoting clean cooking. These and other organizations will be necessary partners in the efforts to promote such new technologies at the local community level. A prominent example outside of the EAP Region is Bangladesh's Grameen Bank, which has made better stoves a focus of its development efforts. In Vietnam, the Institute of Energy long has supported small efforts on improved biomass stoves, and the Women's Union has become involved in new programs as well. Any new institution for knowledge exchange that has ideas for new programs to promote clean cooking can work with such local microfinance organizations and NGOs as partners that can bring in their experience with local issues and problems.

The private sector also is actively involved in the development of technologies to promote clean cooking. At both the international and local levels, private companies are engaged in developing better technologies, participating in the discussions of standards and ratings systems, and setting up distribution and sales campaigns. Although these efforts are at an early stage, this new development should be encouraged through support from international organizations.

Pathway to Implementation

Although some improved stoves are available today, there is still a need for a period of technical assistance to develop locally appropriate options. The basic components that are necessary to move forward and address the problems facing the adoption of these stoves are:

- Ways to encourage or ensure the participation of governments, NGOs, and the private sector
- Assessment of the best methods for finance or subsidies to lower the higher initial costs of stoves
- Improvements in testing protocols and development of standards for products that are reliable, durable, and efficient
- An organization or combination of organizations to take charge of a development issue that straddles many sectors including energy, environment, health, forests, and gender.

Many lessons can be learned across the Region from small energy programs that have succeeded or experienced problems. EAP countries also are at different levels of both awareness of the problems associated with cooking with traditional stoves and their efforts to promote better models. Large countries such as China have been doing basic research on this issue and implementing programs for many years. Smaller countries such as Cambodia have NGO-run programs for charcoal stoves that have achieved a large scale of operation. The way forward will vary by country.

Overall, the EAP Region needs Region-wide capacity to deal with the issues that are common to the EAP countries. These issues include standards; testing protocols; and ways of financing first cost that encourage market development, effective communication strategies, monitoring and evaluation techniques, and market studies to assess the suitability of various stoves. For instance, it would be essential to identify centers at either the Region or the country level that could test the existing improved stoves in the Region according to guidelines that are agreed by the EAP countries. International donor programs will need assurance that the promotion of stoves actually has development impacts and that the stoves are clean, reliable, and safe. A large country such as China has the resources to accomplish almost all of the functions necessary for having successful small energy interventions such as better stoves.

The need for additional research and policy development should go hand in hand with the development of projects that deliver better stoves to the target beneficiaries. Such projects already operate in the EAP countries. The goal is to expand them and implement new ones—perhaps initially on a modest scale—while providing feedback on technology development to manufacturers.

During the first 2–3 years of the program, a period of testing and project assessment is anticipated to ensure that the stoves and other interventions deliver on their promises of reduced pollution and greater energy efficiency. Significant experience in the international community already is available. Centers of excellence in product development and promotion could lead the way toward ever larger programs leading to universal access to better stoves, while pursuing synergies with other programs that promote better lighting and heating systems.

The challenges of promoting clean and efficient cooking practices have been addressed in many regions of the world (World Bank 2011a). Successful programs have been developed in Africa, Latin America, South Asia, as well as East Asia and the Pacific. These programs range from promoting charcoal stoves that have minimal costs to promoting fairly expensive and substantial wood stoves. A substantial body of knowledge also exists on how to promote private sector participation in the development, marketing, and sales of products that improve people's lives. This wide range of experience with a variety of different stove types offers lessons that can be the foundation for the development of new projects or programs.

The way forward comprises several key elements (World Bank 2011a). Such a strategy would require cooperation among government, the private sector, and NGOs qualified to support program dissemination.

- Some form of government institutions or agencies would be needed to facilitate the program.

- Incentives would need to be provided for NGOs, the private sector, or other organizations to address demand-side interventions, principally more efficient stoves.
- Technical development of all types of stoves is still necessary. The leading manufacturers of advanced stoves continue to refine and develop new models.
- Monitoring and evaluation for both improved and advanced stoves are necessary under conditions of actual stove use and to link grants or subsidies to program performance.
- Grants and financing would be necessary to encourage the development of businesses or other organizations to sell or retail stoves. The grants can be channeled through retail markets or organizations that can provide both financing and quality assurance for products.

Financing from international organizations to support the development of clean cooking programs is essential to achieve the elements of this strategy. This financing will require the cooperation of many international organizations to ensure that the program focuses on the commercialization and sustainable scaling-up of projects that support clean and efficient cooking. Consideration could be given to the development of an agency based on the experience of the Water and Sanitation Program (box 5.3).

Outlook for Modern Cooking Solutions

For many years, the health and environmental issues arising from burning biomass energy in primitive stoves or open fires have been on development agencies' list of priorities. Many of the early cooking stoves that were promoted under various national programs were untested and of poor quality. In addition, the difficulties of promoting small products that improved energy efficiency for cooking, lighting, and heating through retail markets or rural development programs are well known.

However, the outlook for modern cooking solutions has changed for the better. Despite past problems, during the last decade, new products have been developed. The model for implementing better stove programs can follow a path similar to that of small, renewable energy systems deployed for those who lack either main grid electricity or modern fuels such as LPG for cooking. The path forward to reach universal access will have to include the promotion of both small appliances provided through retailers or NGOs and the promotion of electricity and LPG by large energy companies that work through network systems or distributors.

The pieces are falling in place for the EAP Region's intensified promotion of clean cooking solutions. New technologies are being developed through the cooperation of private organizations and donors. Climate change funds are increasing their attention, and opening their doors, to financing clean and efficient cooking and ensuring that lack of funding is not an obstacle to the growth of retail businesses that serve the poor. New international alliances have been formed to more actively support and promote clean cookstoves in ways consistent with past successes. The increasing international focus on improving the productivity of rural women and, more generally, on clean and healthier homes is directly related to promoting better cooking practices. The private sector has joined the efforts to develop new products for improving cooking practices and making commitments to focus on the poorest segments of society.

The remaining challenges of scaling up the provision of energy services to the poorest people in the world should not be underestimated. For some time, it will be necessary to support the technical development of products suitable for poor and rural people in EAP. This support will be necessary because the target beneficiaries make up the bottom of the pyramid. They cannot afford to support the technical innovations for the continued development of new products. However, it should be cautioned that many past programs that have focused on the poorest households have promoted low-cost products that were not much better than traditional stoves. With 1 billion people using solid fuels in the Region, better products manufactured on a large scale are essential to reduce costs enough to reach those in the lowest income groups. There also is a need to support the efforts to promote clean cooking being undertaken by microfinance organizations, private companies, government, and NGOs.

Apart from improved cookstoves, it is important to promote the use of LPG, kerosene, biogas, and biofuels in ways that are economically and environmentally sustainable. These fuel options are extremely important and should be actively pursued. However, their reach is limited to those who can afford either the monthly costs or the rather high costs to purchase the initial system.

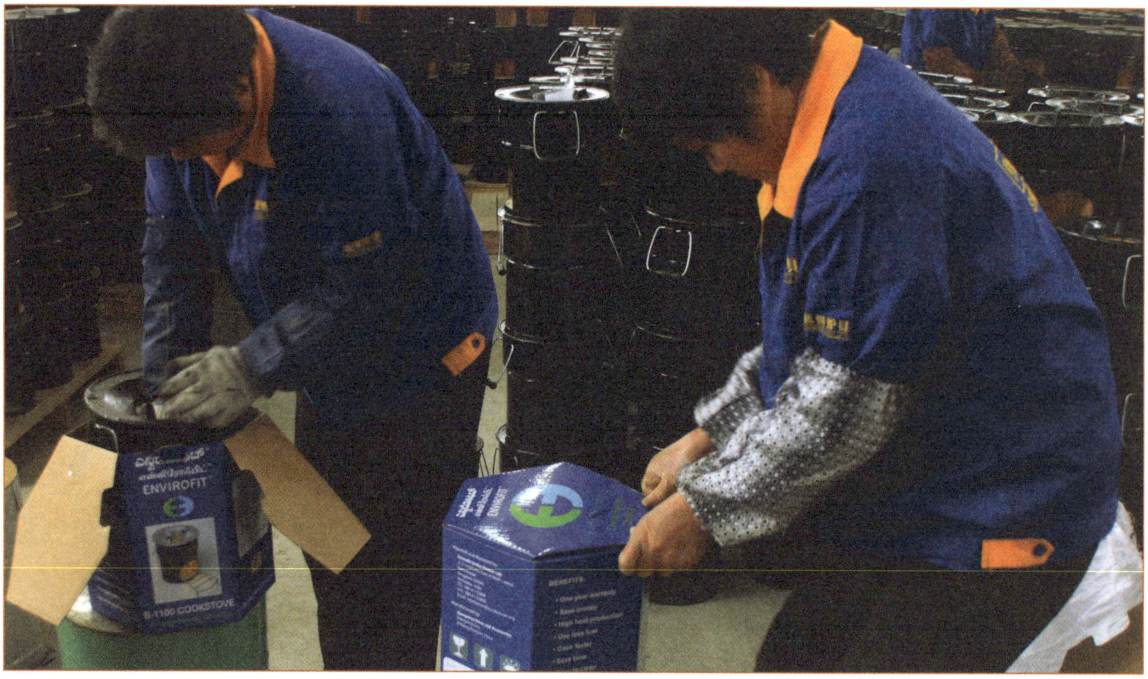

Top: Testing improved cookstoves, China.
Bottom: High-quality, industrially produced improved cookstoves, China.

Photo credits (top and bottom): Ashden Awards.

6

One Goal: Achieving Universal Energy Access in the EAP Region

This report concludes that two separate paths—electricity, and clean and efficient cooking solutions—lead to achieving the one goal of universal access to modern energy in the EAP Region. This chapter discusses the overall strategy to achieve this goal via the two paths.

Both Paths Must Be Implemented

To achieve universal access to modern energy in the Region, there are no trade-offs between the two paths. Both must be implemented. However, in the past, the progress on these two paths has been uneven. Even though many EAP countries have made significant progress toward achieving universal access to electricity, the challenges and strategies differ for each nation (appendix 1). Nevertheless, there is enough international, Regional, and in-country experience to design and successfully implement national plans for universal electricity access within the next two decades. Most important, electrification remains a top priority for governments of EAP countries that have low access rates.

In the past, EAP countries have given more attention to electrification than to access to modern cooking solutions. For example, Lao PDR has provided electricity to approximately 70 percent of its people, whereas less than 10 percent of its people cook with modern fuels or improved stoves that use traditional fuels. Overall, every second household in EAP depends on solid fuels (coal and wood-based biomass) for cooking. Therefore, a major breakthrough is needed to significantly increase access to both modern cooking fuels (natural gas, LPG, and biogas) and improved cookstoves for biomass, particularly in poor and remote rural areas.

Both Paths Will Bring Significant Economic, Social, and Environmental Benefits to All EAP Countries

Benefits of electricity access are high and generally well understood. Electricity can provide the basis for investments in new productive activities and job creation. Conventional power can increase agricultural productivity through motorized water-pumping and value-adding grain-processing facilities. Lighting from grid-based electricity or solar PV systems helps to extend the number of active hours for domestic and social life and small businesses and stores. Even simple solar lanterns improve the quality of life by providing essential lighting and convenient charging of mobile phones, thus facilitating communication and making it easier for farmers and other rural users to receive timely market and weather information.

Recent studies confirm that the benefits of rural electrification are quite high. In a broad review of World Bank projects, the benefits of electricity were found to range from $10–$20 a month per HH and sometimes higher, or up to $1 per kWh. For household lighting alone, the benefit ranged from $0.47–$0.81 per kWh (IEG 2008). In contrast, the life-cycle cost of supplying electricity to rural areas is much less, at $0.15–$0.20/kWh for grid-based rural electrification and up to $0.65/kWh for remote, off-grid systems (World Bank 2007b). Taking into consideration its high potential to improve health, education, and communications, electricity clearly is essential to enable countries to experience both economic and social progress.

Benefits of Access to Clean and Efficient Cooking Solutions Are Important to Achieve the Millennium Development Goals (MDGs) in EAP Countries

The direct benefits of cleaner and more efficient cooking methods arise from the improvement in respiratory health and other illnesses caused by indoor pollution from cooking with solid fuels. When household members are in better health, they are more productive; they miss fewer days of work and school. Moreover, better health reduces medical expenditures, which can be a large portion of expenditures among the poor. In addition, improved cooking solutions directly and principally benefit women and children, who bear the brunt of the physical labor and time expended to collect traditional fuels and the adverse health impacts of the indoor air pollution created by traditional cookstoves.

Reducing the use of traditional fuels in cookstoves also may result in extensive environmental benefits. For example, the reduction in wood collection can revitalize the local environment in some areas that either are densely populated or have poor climatic conditions for growing biomass. In addition, fuelwood that comes from unsustainable local biomass resources actually emits significant incremental quantities of CO_2 into the atmosphere.

Figure 6.1 Investment Requirements for Universal Access to Electricity and Modern Cooking Solutions in EAP, 2011–30

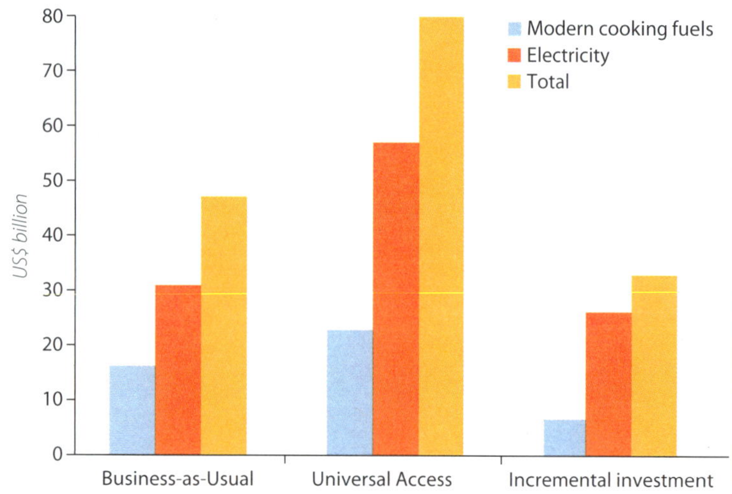

Source: Authors' calculations.

Both Paths Are Affordable but Require Significant Subsidies to the Poorer People

The combined investment for electricity and clean cooking solutions in the Business-as-Usual scenario from 2011–30 is estimated to be $46 billion. The combined investment requirement for the Universal Access scenario over the same period is estimated at US$78 billion. This combined total investment of US$78 billion spread over 2 decades is a small fraction of the Regional GDP,[34] and the incremental investment of US$32 billion is affordable from a Regional perspective (figure 6.1).

However, the annual incremental cost of the electricity path—US$1.3 billion—is an order of magnitude more expensive than the annual incremental cost of the clean cooking path (US$86 million, excluding China). The reason clearly is that supplying electricity inherently is much more capital intensive than providing advanced cookstoves, even though the population benefiting from the latter would be four times more than the number of electricity beneficiaries. Furthermore,

34. At current prices, the combined 2009 GDP of Cambodia, China, Indonesia, Lao PDR, Mongolia, the Philippines, Thailand, and Vietnam was approximately US$6 trillion. Excluding China, GDP was approximately US$1 trillion.

the unit costs of electricity supply rise as service is provided to distant and scattered communities. In contrast, the expenses related to advanced cookstoves are not likely to increase in the same proportion, even for remote communities.

The high cost of the universal electrification program is particularly challenging for poor, low-access countries, for which the incremental investment needs are approximately 1.6 percent of GDP. This high cost calls for significant annual subsidies or concessional financing on the order of US$200 million (approximately 1 percent of their GDP in 2009) to assist the low-access countries to achieve universal access by 2030.

Strategic Framework for Achieving Universal Access to Modern Energy

To achieve universal access to modern energy, both paths need a sustained national commitment and a clear division of roles and responsibilities. Based on the experience in EAP countries and worldwide, a sustained national commitment is a key success factor in reducing energy poverty. Although private sector and nonprofit organizations will have key roles, the goal of universal access will not be achieved without effective and accountable government actions. On the other hand, a government alone cannot efficiently deal with all aspects of expanding access to all consumers. This goal requires concerted action by the public and private sectors, donors, and NGOs; as well as continued learning, sharing experiences, and adapting approaches and instruments to the conditions on the ground.

In each EAP country, significant benefits can arise from designating a single agency or office as responsible for delivering the one goal of universal access to electricity and modern cooking solutions. So far, no EAP country has looked at universal energy access in a holistic manner that includes both electricity and modern cooking solutions. This partial view has led to the relative neglect of providing modern cooking solutions for the large proportion of populations that continues to depend on traditional cooking fuels and methods. Each country government will formulate its overall strategy and targets for universal energy access according to its own strengths and needs. This effort should be given the high profile and urgency that it deserves by designating an agency or a high government office with overall responsibility for the overarching goal of national universal energy access. Under such an arrangement, electricity and cooking will receive commensurate weight at the policymaking and higher implementation levels. This high-level unification of responsibility also will facilitate coordination, both within the two paths (electricity and modern cooking solutions) and with other linked sectors, including environment, health, forests and gender.

Since implementation on the ground would be carried out by a number of official, private, and nonprofit institutions and agencies, the designated responsible agency would determine its roles and responsibilities. In particular, this agency would determine the key parameters of the program. They would include the annual or multiyear targets, sequencing of geographical areas, share of off-grid options in electricity access, share of improved cookstoves among all cookstoves using solid and biomass fuels, and total official funds needed to support the electrification and cooking programs.

The designated agency also would liaise with international and Regional institutions and channel funds and knowledge into the country. Finally, this agency would be responsible to monitor and evaluate progress and recommend any changes arising from lessons learned during implementation or from shifts in national or international circumstances.

The responsible agency should open two action fronts in fighting energy poverty. The first front comprises all urban and many rural consumers in areas in which the sufficiently high density of energy consumption justifies network-based solutions for their energy needs. This first front consists of households who, under the Universal Access scenario, will receive grid-based electricity and LPG/gas/kerosene for cooking (figure 6.2). These consumers generally will be served by

Figure 6.2 Focusing on the "Second Front": Households beyond Reach of Standard Delivery Mechanisms for Modern Energy Access

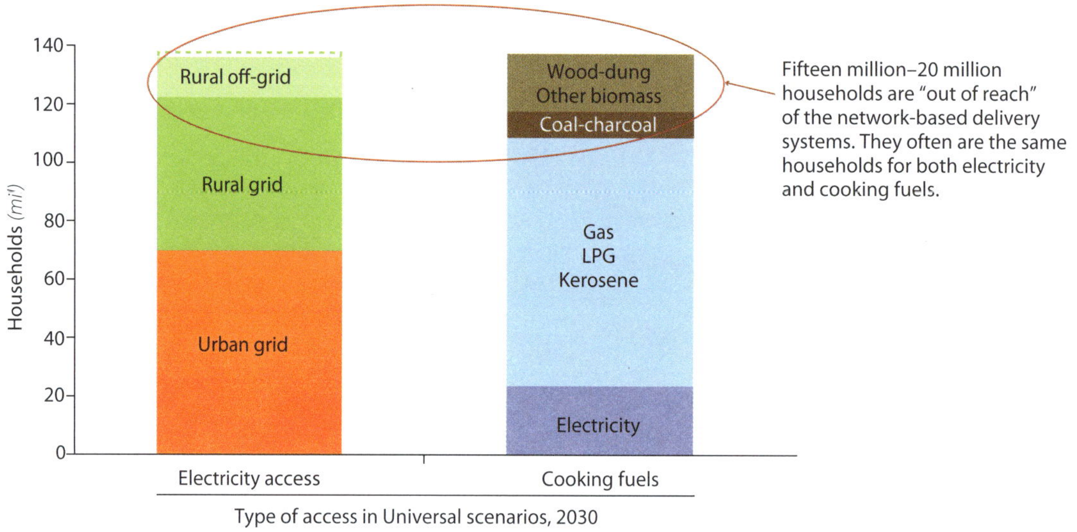

Source: Authors.

existing utilities using network-based approaches. These consumers are designated as the first front of the universal access program in the sense that the technologies and institutions to serve them are reasonably well established in most EAP countries. The second front—comprising people in poor and remote areas—requires a decentralized approach based on distributed energy solutions designed to meet the needs of low-income consumers. Most important, fighting energy poverty on the second front should not be postponed until energy poverty is eradicated on the first front. Both fronts are equally important to achieve universal access to modern energy in a timely manner (by 2030). Therefore, simultaneous action is required on both fronts.

The populations belonging to the two fronts can be viewed in terms of the nature of the products that they will receive and the agencies that will serve them. Under the Universal Access scenario, it is projected that 15 million–20 million EAP HH (75 million–100 million consumers, or 15 percent–20 percent of the total EAP Region population, excluding China) cannot be served by extending the power grid or by petroleum company distribution networks. For electricity, these consumers will have to be served by off-grid options such as independent microgrids, solar home systems (SHS), and other distributed energy solutions. For cooking, these consumers will have to be served by improved cookstoves that burn traditional biomass fuels, coal, or charcoal. Either these consumers are located in remote, rural areas that the existing agencies will not be able to reach in a timely manner, or the agencies do not offer products suitable for these consumers. Generally, these consumers have lower incomes and low levels of energy consumption. These characteristics make the network-based solutions economically unviable and likely unsustainable, even if significant subsidies are provided.

Strategic Approaches for the Two Action Fronts Will Differ

On the first action front, the approach will be to extend the reach of large energy suppliers using network-based solutions. The second front, however, requires a decentralized approach based on distributed energy solutions designed to meet the needs of poor consumers, often in remote areas.

Action Plan for the First Front

In most EAP countries, large energy companies already provide modern energy services using network-based solutions. For instance, in almost all urban areas, households expect a grid-based electricity connection and fairly high levels of reliability of electricity supply. Likewise, most urban areas host well-established national or private petroleum distribution companies that sell LPG and kerosene. These companies, suitably strengthened, should have the responsibility of connecting all urban HH to electricity and shifting them to modern cooking fuels.

In rural areas—with the exception of China, Thailand, and Vietnam—programs will be necessary to strengthen grid-based rural electrification cost effectively. This responsibility can be entrusted to the urban-based utility, as is common, or to a separate rural electrification agency. Regardless, it is essential that the responsible agency has sound governance systems, good management, healthy finances, adequate technical capacity, and a corporate culture of cutting costs. Cost reduction is essential to keep the total subsidy requirements manageable (chapters 2 and 3).

To expand petroleum-based cooking fuels to rural areas, in most cases, it would be natural for the urban-based agencies to be responsible for increasing access through expanding their supply chains to rural areas. Such is the case in Thailand, which has EAP's highest level of LPG use in rural areas. This success was accomplished through subsidizing regional LPG storage depots that could be used by multiple private sector companies to serve people in rural areas. Since circumstances vary significantly among countries, care must be taken to ensure that rural consumers are not neglected or overlooked by the service providers and other service entities, which may find it easier to continue with their mainly urban customers.

With respect to electricity, the existing utilities will need to take special steps to serve poorer consumers who may not be able to pay the full cost, particularly the upfront capital cost, of the service that they receive through network-based solutions. It may be necessary for a tariff-setting authority, such as an energy regulator, to develop cross-subsidy schemes that would make the service affordable to these consumers. It also may be useful for the utilities to develop links with community groups and NGOs that are familiar with these consumers and can facilitate specifically targeted schemes to serve them. This arrangement could work in slum areas in which such organizations could be a voice for the service concerns of low-income populations.

Action Plan for the Second Front

The expansion of modern energy services to those who are not within reach of either the grid electricity systems or petroleum fuels for cooking will require a more decentralized approach and the use of distributed energy systems. The number of consumers in this group will be fewer than those being served by the major energy carriers using network-based solutions (figure 6.2). For this reason, unless these underserved consumers are specifically targeted, governments may tend to neglect them due to their primary focus on expanding national electricity and other energy networks.

In future, governments of low-energy-access countries must make a special effort to serve poor and remote consumers in parallel with their efforts to expand access to consumers on the first front. In particular, it is necessary to develop suitable technical products and workable financial schemes to meet the needs of consumers on the second front. Nevertheless, this special focus on poor and remote populations definitely should not lead to reduced funds and support for the expansion of energy services to consumers on the first front, who form the bulk of the underserved and who also are largely poor.

Since providing universal energy access to poor and remote households is not easy and has not been a mainstay of energy policies in most countries, these countries now must face certain institutional, technical, and financing issues.

Institutional issues. The core of a second-front strategy is that each EAP country either designates an existing agency, or creates a new one, to take overall responsibility for access to energy—electricity and modern cooking solutions—for the poorest or most inaccessible households. This agency could contain separate implementing units or programs for electricity and cooking solutions. Alternately, if implementation capacity is limited, the agency could be organized as a single unit. The structure of this agency could vary by country. What is important is that it be held accountable to develop programs that reach the poorest households with electricity and more modern ways of cooking.

Decentralized systems of rural electrification traditionally have not been seen as a natural extension of grid-based RE. Even when existing electricity companies have been designated with the responsibility to provide off-grid electricity, these companies have not always been suited to carry out such programs. Nevertheless, a grid-based RE agency could be suitable for off-grid electrification if this role were given adequate attention in the overall mission of the agency. If an electricity company were designated as the national electrification agency, it also could consider an institutional mechanism, such as a dedicated well-staffed subsidiary, to implement the off-grid electrification program.

Alternately, a separate agency could be used for off-grid RE provided that it could be established quickly and could attract suitably qualified staff. One advantage of a separate off-grid agency is that it could exploit synergy in promoting off-grid electricity and clean cooking solutions since both involve similar methods for commercialization of smaller scale energy products and services. Regardless of the choice of agency, it is essential for governments to ensure that off-grid RE is undertaken expeditiously and in parallel with grid-based RE.

The development and dissemination of ad--vanced cookstoves have no synergy with the expansion of LPG networks in rural areas due to significant differences in the products, costs, and target groups whom LPG and improved cookstoves will serve. The petroleum companies that will be important for promoting LPG in both urban and rural areas have no comparative advantage in the commercialization of improved cookstoves. In general, a separate implementing agency for improved cookstove programs is warranted. However, if an existing agency is selected (such as the off-grid electrification agency), it must ensure that an institutional mechanism exists within the agency to guarantee that cookstoves are not neglected.

Private firms, community groups, and NGOs all will have important roles to play in serving the consumers on the second front. Profit-oriented firms and NGOs, such as Grameen Shakti in Bangladesh, have been involved in disseminating solar home systems and solar lanterns in the rural areas of many countries. It also is common for community groups, NGOs, and nonprofit-oriented firms to own and operate village-level electricity microgrids. Within the EAP Region and in most developing countries, it has been common for community groups and NGOs to take the lead in the dissemination of improved cookstoves. Thus, all groups have the potential to participate in the provision of decentralized/distributed electricity services and improved cookstoves, with their roles depending on the situation in each country, or region of a particular country. It will be the responsibility of the agency responsible for the poor and remote (second-front) consumers to assess the circumstances and assess the roles of these groups.

Technical issues. Although the products for both decentralized HH electricity and clean cooking solutions are fairly well developed, they continue to evolve and improve at a relatively fast pace. As a result, although good systems are available, this area is likely to remain dynamic in the coming years, and many current solutions will be replaced with improved versions.

In the past, solar systems commonly were configured as "solar home systems (SHS)," which provided enough electricity for multiple lights and other uses, such as a television. While SHS are a well-established product in EAP, they are evolving. New types of lights (light-emitting diodes, or LEDs) are being introduced. Their lower energy use has to be balanced against their higher initial cost and the lack of field experience with them.

The use of conventional appliances adds to the convenience and flexibility of SHS but in the field may introduce another element of cost and possibility of technical failure. Finally, more efficient solar panels are being developed.

As mentioned above, in recent years, "solar lanterns" have emerged as a new product that provides enough electricity for a single light and charging a cell phone. They are a promising option to meet the basic electricity needs of second-front consumers. However, solar lanterns are evolving rapidly—much faster than SHS—in technology and capacity. Practically, to select solar lanterns that are suitable, cost-effective, and not outdated will present a challenge for each EAP country. Thus, programs to promote decentralized electricity systems should be flexible and open to innovation.

Over the last 5 years, cookstoves also have evolved through many innovative designs. A wide variety of manufacturers are working on additional improvements. The future looks very promising, but no clear designs can be considered fully developed or mature. At present, governments need to develop technical standards for improved cookstoves, and introduce type testing and certification procedures based on these standards. There also is a need to launch a consumer awareness program to educate the target beneficiaries about the features, benefits, and limitations of the new products. This situation resembles the context that prevailed for SHS in developing countries 10–15 years ago. A strong need exists to initiate and accelerate resolving these cookstove issues. However, at present, few EAP countries are equipped to undertake this task.

Financial issues. A major financial issue for SHS is how consumers should pay for them. One option, often called the vendor model, is that consumers buy the systems using capital cost subsidies and credit, as is succeeding in Bangladesh, China, and Sri Lanka. However, this model faces some limitations when applied to solar lanterns. At the current prices of US$20–$40 per lantern, subsidies are needed to make them affordable to poor households. Most programs that finance the development and marketing costs of solar lanterns do not use subsidies. The reason is that, if the capital cost is subsidized, poor consumers can easily resell them in the market to better-off households who may be willing to buy a supplemental source of electricity. How to finance affordable small lighting systems will be an issue in scaling up any major new programs.

A second option for financing small electricity systems is that consumers pay a monthly electricity bill to a utility service provider, which owns and maintains the solar systems installed in individual homes. Often called the energy service company (ESCO) model, this type of program has become popular for reaching the poorest households in Latin America. The ESCO arrangement makes it easier to target the number of HH to be served during a particular period than does the vendor model. However, the ESCO model places some revenue risks on the utility because the financial viability of the service company depends on its ability to collect relatively small monthly bills over a long period and to adjust the price charged when costs rise due to overall inflation. Although Latin America has had some success with the ESCO model, it has not yet been very popular in EAP. Thus, each EAP country will have to make the choice regarding how to reach out and make services affordable to poor and remote populations.

Similarly, financing improved cookstoves (which are comparable to solar lanterns in cost) has been based on a wide variety of different models. It is now generally accepted that it is preferable to reduce cookstoves' upfront cost through support for technical development, marketing, and retail services; and to sell the stoves through the marketplace without direct consumer subsidies except perhaps rebates. In this way, the stoves that are purchased and appreciated by consumers could be included in more general development programs as well. Moreover, once quality stoves are available in the marketplace and valued by consumers, microfinance organizations could provide consumer loans for their purchase.

Regional Entity for Universal Energy Access

The institutional, technical, and financial issues discussed above are similar across the EAP countries and across South Asian countries as well. However, most EAP countries are developing their own programs for achieving universal energy access in relative isolation. While each country will tackle these three issues taking into consideration its own specific circumstances and level of development, it would be more cost-effective and faster if countries could learn from one another and from international best practices through a Regional *entity*, which would facilitate knowledge sharing and capacity building. This Regional entity also would be a focal point to channel the assistance of various interest groups, such as donors and NGOs, who are interested in improving the quality of lives of poor households.

It would be essential that any new Regional entity have the endorsement of the beneficiary countries. Hence, this report recommends that this issue be discussed with the EAP countries concerned and that additional steps toward establishing such an entity be taken in consultation with them. A Regional entity on universal energy access dedicated to working with governments and their implementing units and committed to sharing knowledge with donors, financing agencies, NGOs, and the private sector could play an important role. Such an entity could keep track of energy access programs in the Region, facilitate the implementation of more effective courses of action, and ensure that the issue of universal energy access is given its deserved attention and public visibility.

An alternate approach might be to establish a Regional or international *program* to deal directly with these issues. The Water and Sanitation Program (WSP), which deals directly with related issues at the HH level, employs this approach. WSP places dedicated staff in the Region. They are supported by a central group, who deal specifically with the necessary government agencies and related programs. Regardless of the Regional approach, there needs to be a credible international entity that ensures that the issue of universal energy access gets the attention that it deserves at the Regional level.

APPENDIX 1

Electricity Access: Selected Country Briefs

A1.1 Cambodia

With the exception of Myanmar and most Pacific Island Countries (PICs), Cambodia has the lowest electrification rate in EAP. In 2009 approximately only 26 percent of its 2.8 million HH were connected to several isolated grids. Cambodia's electrification levels stand in stark contrast to those of its neighbors: Lao PDR (70 percent electricity access), Thailand (99 percent), and Vietnam (95 percent). Access to electricity within Cambodia is highly uneven. Approximately 87 percent of the urban population is covered, compared to only 13 percent of the rural population. The government has set an ambitious target of providing some form of electricity, including access to

Population *(2008) (mil)*	13.4
Rural population *(% total population, 2008)*	80
Population density *(people/sq km)*	82
Land area *(sq km)*	181,035
GNI per capita *(Atlas Method: Current US$, 2008)*	610
Access to modern cooking fuels	7.9
Urban *(% HH, 2008)*	37.3
Rural *(% HH, 2008)*	1.5
Electricity access, national *(% HH, 2009)*	26
Urban *(% HH, 2008)*	87
Rural *(% HH, 2008)*	13
No. of people w/o access to electricity *(2009) (mil)*	11.3
Population served by off-grid sources (minigrids and HH systems) *(%)*	4
Electricity access target and year *(% HH)*	70; 2030
Electric power consumption *(kWh per capita, 2007)*	124
Installed capacity *(MW 2008)*	386
Thermal	373
Hydro	13
Other renewable	—
Electricity net generation *(bil kWh)*	1.38
Distribution losses *(% net generation)*	10.5
CO_2 emissions *(M/T per capita, 2007)*	0.3
Indicative residential electricity tariffs for rural consumers *(2011)*	USc23.5/kWh. Ranges 670–1220 Riel (*Source:* Electricity Authority of Cambodia 2009)
Key institutions in the electricity sector	Ministry of Industry, Mines and Energy (MIME) Electricité du Cambodge (EdC) Independent regulator: Electricity Authority of Cambodia (EAC) Rural Electricity Enterprises (REEs)

Sources: World Bank 2010e; www.eia.doe.gov; IEA 2010; WHO and UNDP 2009.

Cambodia is bounded by Thailand on the north, Lao PDR on the east, Vietnam on the southeast, and the Gulf of Thailand and Thailand on the west. Much of the country's area consists of rolling plains. Dominant features are the large, almost centrally located Tonle Sap (Great Lake) and the Mekong River, which traverses the country from north to south.

Table A1.1.1 Cambodia: Scenario Analysis for Universal Electricity Access by 2030

	Total investment needs *(US$ mil)*		Incremental needs for universal access *(US$ mil)*			
Period	**Business-as-Usual scenario**	**Universal Access scenario**	**Total**	**Urban grid**	**Rural grid**	**Off-grid**
2011–20	315	1,446	1,131	80	801	250
2021–30	355	2,337	1,982	—	1,602	380
2011–30	**669**	**3,783**	**3,113**	**80**	**2,403**	**630**
		Annual requirements:	156	4	120	31

Note: No. of HH without electricity in Business-as-Usual scenario in 2030: 2.4 million (60% of population).

minigrid and off-grid electricity, to 70 percent of rural HH by 2030.

CHALLENGES

- Government's leadership and "visible hand" in promoting energy access yet to materialize
- Sector structure fragmented and of low capacity
- Coherent and workable nationwide plan for improving access yet to be developed
- Institutional and financial frameworks for off-grid electrification weak or absent

- High costs of power supply major barrier to scaling up access
- National program to improve access in need of large-scale concessionary financing

STRATEGIES

- Under the leadership of Electricité du Cambodge (EDC), the official power utility, jump-start the national electrification scale-up in a systematic and programmatic mode
- Move toward sector-wide approach (SWAp) and programmatic framework
- Build a comprehensive and credible plan for grid rollout and its off-grid complement
- Rationalize sector structure, especially role of Rural Electricity Enterprises (REEs)
- Develop sustainable financing platform and subsidy and tariff framework
- Revisit centralized planning and implementation of rural electrification and make necessary changes to advance RE program

CHALLENGES

Government's leadership and "visible hand" in promoting energy access have yet to materialize. Cambodia has set a target of electrifying 100 percent of its villages and 70 percent of its rural HH by 2030. However, there is little evidence of government leadership and "visible hand" advancing this commitment. The targets have not yet been backed by an enabling policy and financial commitment. There does not appear to be "buy-in" at the cabinet level or consensus by the key stakeholders—EDC, EAC, REEs—or the private sector. Overall, the lack of a "champion" with a clear vision has resulted in a policy vacuum and stagnation in the electricity access effort.

Sector structure is fragmented and of low capacity. Electricité du Cambodge (EDC), the national power utility, is structured as a vertically integrated national power company that serves Phnom Penh area and some provincial centers. Additionally, 249 regulated private sector entities called Rural Electricity Enterprises (REEs) are engaged in transmission, generation, and/or distribution of electricity, mainly through diesel-based minigrids throughout the country. Most REEs are in poor technical and financial condition. EDC is much stronger in technical capacity and financial terms than the other entities. Thus, EDC is the only credible sector institution to anchor the planning and implementation of a national grid program for nationwide electrification.

The Electricity Authority of Cambodia (EAC) is the independent regulator. EAC is responsible for licensing electric power suppliers, managing tariffs and fees, and generally regulating the economic environment of power production.

The current structural and transactional framework rules for key sector institutions—EDC, REEs, EAC, and the private sector—are not conducive to national program planning and implementation, even with a ready, willing, and able EDC. There needs to be a clear policy that clarifies the respective roles and scope of sector institutions in the current context and a roadmap for how they would evolve in the future. The principles that should govern the transactions and interface between institutions and regulatory oversight need to be clarified and harmonized into a well-coordinated and functioning whole.

A coherent and workable nationwide plan for improving access has yet to be developed. Cambodia has yet to systematically develop and implement a set of socioeconomic criteria, similar to those used by Lao PDR, Thailand, and Vietnam, to provide a clear, technical basis for network expansion. Consequently, a national grid rollout plan that is analytically sound, comprehensive, spatially representative, and current has yet to be prepared. While Cambodia does use geographic information systems (GIS) in its current power system planning, the technology needs to be upgraded and training provided to raise capacity to acceptable standards.

High cost of power supply is a major barrier to scaling up access. Lack of power supply at reasonable cost remains a major barrier to scaling up access. EDC has had high generation costs due to high diesel prices, small-scale operation, low technical capacity, weak management, and non-transparent investments in generation and transmission. Consequently, the current tariff rates for EDC (US$0.09–0.23/kWh in Phnom Penh) are among the highest in the world.

Of the 249 REEs, approximately 226 are engaged in distribution and retailing. Fifty percent of them offer 24-hour service; the rest offer only 4–12 hours' supply per day. Most REEs have a generation capacity in the 1–10MW range and typically serve 500–2,000 customers.

The REEs have the same high-cost features as does EDC. Their costs are even higher than EDCs because they use expensive diesel fuel. Correspondingly, the REE tariffs, in the range US$0.20–1.00 per kWh, are higher than EDC tariffs. Despite their higher fees, most REEs are financially weak. In addition to low technical capacity and weak management, they have little possibility of expanding their operations and gaining economies of scale.

Tariffs are regulated by EAC and are determined separately for each REE licensee. A survey estimated that rural HH can afford to pay approximately US$7.50 monthly for electricity supplies (AECOM 2009). This amount is equivalent to a tariff of approximately US0.15c/kWh, compared to a cost of supply of US.23c/kWh or more. The difference needs to be made up from other sources including cross-subsidies from urban customers, subsidies from the government, and concessionary financing from donors.

A national program for improving access needs large-scale concessionary financing. The government's target of extending access to 70 percent of all HH by 2030 translates to connecting approximately 100,000 rural HH annually between now and 2030, or 2.2 million rural HH in all. The total cost of this electrification program (excluding generation and transmission) is estimated at US$2.6 billion. Given its current large budget deficit, the government is unlikely to be able to make any significant financial contribution to the costs of a rural electrification program from tax revenues.

Institutional and financial frameworks for off-grid electrification are weak or absent. Cambodia has no clear institutional focal point that systematically will pursue off-grid electrification. A semi-autonomous Rural Electrification Fund (REF) has been set up with the mission of "rural electrification," with representation from MIME (Ministry of Industry, Mines and Energy). REF provides a household connection subsidy of US$45/HH for new connections by REEs within their service areas. This subsidy has resulted in over 42,000 additional HH being connected. REF received its initial capitalization from a GEF project but has no provision for reflows back into the fund, and no sector or fiscal funds have been earmarked for fresh capital inflows. REF has the responsibility to promote the solar home system (SHS) program and microhydro grids of approximately below 1MW capacity each. However, REF currently lacks the technical expertise to design either the SHS program and subsidy scheme or the microhydro program.

Strategies

Under EDC's leadership, jump-start the national electrification program in a systematic and programmatic mode. Good-practice experience confirms that achieving high national electrification coverage efficiently, effectively, and quickly requires a well-orchestrated and well-managed national grid rollout program implemented by a committed national utility that has the leadership, capacity, autonomy, and means to carry out the task. A national grid extension led by EDC should be the main delivery vehicle of the national electrification program. EDC should be the principal implementing agent of the government for planning and securing adequate and affordable bulk supply. The authority should lead the national grid and MV network rollout as well as the LV network development and retailing in areas in which no REEs are operating or are not likely to operate in the future.

At best, the private sector and off-grid service delivery can play an important and complementary niche role but not as the main organizing instruments of national electrification. The efficient and better performing REEs can play an effective role if they are recast as buyers of cost-competitive bulk power supply and manage the distribution and retailing business in their respective service areas.

Move toward a sector-wide approach (SWAp) and programmatic framework. The investments required to fund Cambodia's national electrification program over the next 20 years are huge. Consequently, Cambodia should mobilize sufficient ongoing financing. Doing so can be achieved most effectively within a programmatic framework, rather than by using a fragmented, project-by-project, and donor-by-donor approach. The SWAp would be led by the government of Cambodia and seek to rally partner engagement in alignment with a common sector-wide investment program and implementation and financing plan ("Sector-Wide Prospectus").

The key building block of a sector-wide prospectus is a comprehensive, consistent, and credible strategic-level grid rollout implementation, investment, and financing plan for the national electrification program rollout and its off-grid complement for 2012–30. The underlying basis of this plan should be a GIS-based, least-cost spatial grid rollout planning platform and model framework, which also would determine the off-grid complement.

Currently, 93 percent of energy generated by licensees in Cambodia is based on expensive diesel fuel. The country is building several hydropower plants and a coal-fired plant, which are expected to reduce the share of oil-fired generation and thus reduce average power costs. A national transmission network is being developed. It is taking the form of a number of HV transmission lines connecting to the neighboring countries of Thailand and Vietnam, which will be followed by cross-border connections to Lao PDR. This network will enable the import of lower-priced power. Over time, these lines are expected to be extended to create a single interconnected national grid.

Build a comprehensive and credible plan for grid rollout and its off-grid complement. A key step is to build a strategic and credible grid rollout plan, including implementation, investment, and financing details for the grid and off-grid components (2012–30). The anchor for such a plan should be a GIS-based, least-cost spatial grid rollout planning platform and model framework, which in turn would determine the off-grid complement. An illustration of such an exercise is described in box 2.5.

Rationalize sector structure, especially the roles of REEs. A well-coordinated and harmonized functioning sector should be established to cover current and future licensees vis-a-vis EDC as well as to define principles that govern interface issues and transactions across key sector institutions and regulatory oversight. A suggested first step would be a rapid assessment of today's REE sector performance and capacity along selected technical, commercial, and other key indicators of performance. This assessment would help point the way toward a workable and effective framework that would be politically acceptable while promoting efficiency. Specific steps may include, first, rationalization including consolidation of one or more REEs when and where justified. A second step could be to select better-performing REEs to be strengthened as sector-retailing agents with access to bulk supply from EDC grid bulk delivery transfer point(s).

Develop a sustainable subsidy and tariff framework. No country has achieved high electrification rates without ongoing concessionary finance being made available to the implementing agents. This issue is particularly relevant to Cambodia given its low income status and limited internal sources of revenue. The bulk of electrification needs in Cambodia are in its rural areas. Outside of the major urban areas, electrification is not commercially viable. Nevertheless, the electrification program cannot be sustained over the duration without a financially viable utility, which recovers, at a minimum, all recurrent costs through an appropriate tariff and subsidy framework.

Revisit the centralized planning and implementation of rural electrification and make the necessary changes to advance the RE program. Currently, planning and implementation of the scope, scale, and areas of RE expansion are centralized. A bottom-up approach from the subnational to the national level would be more effective. Such decentralization could pose challenges to EDC at the subnational, if not the national, level. To improve RE planning and implementation, it would be necessary to devolve accountability to the subnational level.

A1.2 Indonesia

Indonesia is characterized by a high population density in the Java-Bali region and progressively lower densities in other regions. As of 2008, 55 percent of Indonesia's 227 million people lived on Java-Bali (among the most densely populated areas in the world). The remainder was spread across Sumatra (19 percent), Sulawesi (7 percent), Kalimantan (5 percent), Nusa Tengara and Maluku (4 percent), and Papua (10 percent). Starting from a very low base of 2 percent national electricity access in the late 1970s, Indonesia's national electricity company, PLN, set a scorching pace for electrification. This growth occurred especially in the decade starting in late 1980s and achieved national access of at least of 65 percent by 1999. By the same year, PLN was connecting over 1 million rural HH per year. PLN was able to achieve this remarkable success by steadfastly strengthening its implementation capacity for planning, design, procurement, and construction services logistics,

Population *(2008) (mil)*	227.4
Rural population *(% total population, 2008)*	47
Population density *(people/sq km)*	125
Land area *(sq km)*	1,904,600
GNI per capita *(Atlas Method: Current US$, 2008)*	US$2,010
Access to modern cooking fuels	46
Urban *(% HH, 2008)*	77
Rural *(% HH, 2008)*	22
Electricity access, national *(% HH, 2009)*	65
Urban *(% HH, 2008)*	97
Rural *(% HH, 2008)*	32
No. of people w/o access to electricity *(2009) (mil)*	81.6
Population served by off-grid sources (minigrids and HH systems) *(%)*	<7
Electricity access target and year *(% HH)*	95–100, 2025
Electric power consumption *(kWh per capita, 2007)*	566
Installed capacity *(MW 2008)*	27,801
Thermal	22.000
Hydro	4,869
Other renewable	932
Electricity net generation *(bil kWh)*	141.2
Distribution losses *(% net generation)*	10.7
CO_2 emissions *(M/T per capita, 2007)*	1.8
Indicative residential electricity tariffs for rural consumers *(2011)*	USc1.9/ kWh under 50 kWh (*Source:* www.pln.co.id)
Key institutions for electricity sector	Ministry of Energy and Mineral Resources Perusahaan Listrik Negara (PLN): State-owned utility

Sources: World Bank 2010e; www.eia.doe.gov; IEA 2010; WHO and UNDP 2009.

Indonesia is an archipelagic country extending 5,120 km from east to west and 1,760 km from north to south. It encompasses an estimated 17,508 islands, only 6,000 of which are inhabited. It comprises 5 main islands: Sumatra, Java, Kalimantan, Sulawesi, and New Guinea; 2 major archipelagos: Nusa Tenggara and the Maluku Islands; and 60 smaller archipelagos.

Table A1.2.1 Indonesia: Scenario Analysis for Universal Electricity Access by 2030

	Total investment needs *(US$ mil)*		**Incremental needs for universal access** *(US$ mil)*			
Period	**Business-as-Usual scenario**	**Universal Access scenario**	**Total**	**Urban grid**	**Rural grid**	**Off-grid**
2011–20	9,684	10,154	470	470	—	—
2021–30	11,078	20,572	9,494	—	7,655	1,839
2011–30	**20,762**	**30,726**	**9,964**	**470**	**7,655**	**1,839**
		Annual requirements:	498	24	383	92

Note: No. of HH without electricity in Business-as-Usual scenario in 2030: 7.5 million (11% of total population).

as well as its on-the-ground implementation of MV and LV networks. Following the macroeconomic crisis that began in 1997 and subsequent organizational changes within PLN, this capacity has deteriorated significantly. The national electricity access rate remains at approximately 65 percent because the rate of increase in electricity access has barely kept pace with population growth.

Introduction and Sector Context: Electricity Access

Aftermath of financial crisis. In 1997 the rapid progress of the grid-based electrification rollout program was abruptly and unexpectedly interrupted by the East Asian financial crisis, resulting in the insolvency of PLN. PLN had huge financial commitments denominated in US dollars under the power purchase agreements with a number of independent power providers (IPPs). The value of these commitments increased several fold due to the huge devaluation of the rupiah, whereas PLN's tariffs, denominated in rupiahs, remained unchanged. In 2001 PLN disbanded the rural electrification division at its headquarters and curtailed most of its rural electrification program. The electrification program has not yet regained speed.

PLN's poor financial health remains a critical issue. A key barrier to electricity access scale-up had emerged pre-crisis and has yet to be addressed satisfactorily and sustainably. Specifically, in the years immediately preceding the financial crisis, PLN's fast-paced national electricity access program had confronted a rapidly rising cost-of-service structure for network extensions and had increased expectations for new connections, including an oil-dominant generation fuel mix, especially outside Java-Bali. By the onset of the financial crisis, the outside-Java operations were losing money in most regions and required increasing and unsustainable levels of cross-subsidies from their Java-Bali operations. On the other hand, PLN's overall electricity sales revenue growth was (and remains) constrained by the level and structure of the uniform national tariff regime allowable for PLN by the Government of Indonesia (GOI).

Government initiatives underway: A good start but unfinished sector reform agenda. The government recognizes that the financial health of the electricity sector is key to Indonesia's growth and developmental aspirations. Over the last decade, the GOI has introduced several initiatives relating to critical pricing, subsidy, and sector reform issues. Noteworthy initiatives are aimed at PLN's decentralizing, unbundling, corporatizing, and restructuring, as well as attracting private sector participation, especially in generation:

- ***Regional business units created to better manage PLN operations and track costs, which vary enormously by region.*** A big first step toward region-specific costing is moving toward tariff and subsidy differentiation and efficient targeting.
- ***2003 State Enterprises Law introduced PSO policy.*** This law empowers GOI to impose a public service obligation (PSO) on state-owned enterprises (SOEs). The law is accompanied by compensation from state budgets for full associated costs plus a margin.
- ***2009 Electricity Law.*** This law opens a door to introduce innovative solutions to expand access. For example, it allows PLN to create subsidiaries that operate independently in decisionmaking and establishing locally cost-based tariffs.
- ***Tariff adjustments.*** GOI recently approved a 10 percent average increase in the national uniform tariff as an initial step toward better alignment with the average cost of service.

CHALLENGES

- Ensure adequate and cost-competitive generation supply to keep pace with robust economic growth
- Extend electricity for lighting to over 80 million Indonesians (33 percent of the population). Of these, over 50 percent are outside Java-Bali
- PLN's financial health remains precarious and a "deal-breaker" for scaling up electricity access outside Java-Bali and for overall sector development

STRATEGIES

- Re-energize the national electrification program implementation with special emphasis outside Java-Bali
- Refocus the public service obligation (PSO) subsidy financing mechanism framework to catalyze electrification programs outside Java-Bali within a programmatic framework
- Set a few good examples in a few promising regions (*Wilayahs*) outside Java-Bali

Challenges

Ensuring adequate and cost-competitive generation supply to keep pace with robust economic growth. Economic growth has returned, resulting in electricity demand growth averaging over 7 percent annually in recent years. GOI's target is to have 55,000 MW in generating capacity online by 2015, compared with approximately 30,000 MW, including PLN and IPPs, installed today.

Its oil-dominant generating mix results in a high cost of service for PLN. Over the medium term, Indonesia has several options by which to invest in cost-competitive energy generation alternatives as well as green energy sources. These options include geothermal, hydro, and larger scale grid-connected solar PV.

The investment costs for an expansion program of this scale and scope, sustained over the medium term, are daunting. Inducing sufficient private sector participation until the related sector reforms are sufficiently advanced remains a challenge. Similarly, the cost to PLN of power purchase obligations needs to be established transparently and competitively. To ensure PLN's ongoing financial health, it also is crucial that these costs be integrated with GOI's tariff policy and PSO subsidy design and financing mechanism.

Over 80 million Indonesians, or 33 percent of the population, remain in the dark; of these, over 50 percent are outside Java-Bali. The government has set ambitious targets of achieving an overall access of 90 percent by 2020 and near-universal access by 2025 across most of Indonesia. To meet these goals will require adding approximately 2 million connections per year from today. In reality, PLN is connecting approximately only 1 million customers annually today—mostly urban and on Java-Bali—and is challenged to ramp up its implementation rate.

Investment costs for the medium and low voltage network extensions required for the access scale-up program are estimated at US$1.3 billion/year through 2025. This estimate assumes widespread deployment of low-cost appropriate network designs and equipment standards as well as cost-conscious construction practices. While physical implementation capacity is a factor, it can be addressed readily.

Figure A1.2.1 Electricity Access in Indonesia: Urban versus Rural, 2009

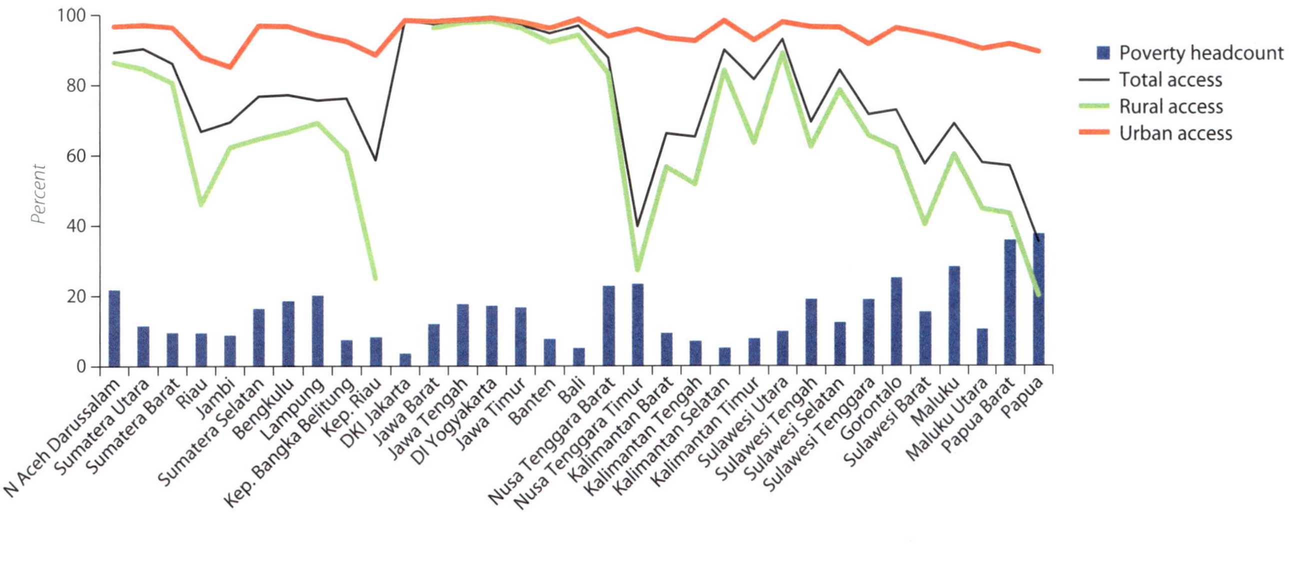

Source: Government of Indonesia.

A significant challenge is to mainstream cost-conscious network design and equipment standards and specifications appropriate for lower density, rural area network extension, especially outside Java-Bali. These features are well proven and widely deployed elsewhere. They help minimize the national subsidy and financing burdens imposed by a rapidly rising cost structure typically faced by the country in its grid extension programs, especially outside Java-Bali.

PLN's financial health remains precarious and a "deal-breaker" for scaling up electricity access outside Java-Bali and for overall sector development. The GOI imposes a public service obligation (PSO) on PLN to connect and supply electricity to all Indonesians. At the same time, the government regulates the average level and structure of retail tariffs that PLN is permitted to charge nationally on a uniform basis. This policy has resulted in PLN facing a tight financial situation despite substantial subsidies to PLN via GOI's PSO financing mechanism towards capital and operating costs.[35]

Due to the importance attached to ensuring the financial health of SOEs tasked with public service obligations, government has opted to progressively refine procedures and processes through piloting their implementation on the most important public service obligation (PSO) providers, notably PLN and PERTAMINA. Considerable progress already has been made in the short period since Law 19/2003 was enacted and Government Regulation (GR) 45/2005 was issued. The government has developed Infrastructure Policy Package maps that identify a set of actions designed to refine the policy framework for subsidy and PSO management in sectors including power.

In summary, attracting qualified IPPs in sufficient numbers and enabling PLN to play its due roles in sector expansion in a timely, efficient, and sustainable manner hinges on effectively addressing the key outstanding pricing and subsidy policy issues. Moreover, the PSO financing mechanism can be more effective in scaling up the access agenda by calibrating its structure and design, and targeting appropriately. Additionally, for sector development to keep pace with robust and sustained economic growth at least cost, attracting private sector participation requires pressing ahead with the enabling reforms. These reforms comprise decentralizing PLN; helping regions to create fully functioning business units with established cost structures; and advancing reforms to encourage transparent and competitive IPP participation.

Strategies

Re-energize the national electrification program implementation with special emphasis outside Java-Bali. Large areas of southern Sumatra, much of Southern Sulawesi, and some coastal city areas of Kalimantan in particular offer relatively high economic and demographic density, so are appropriate for a medium-sized grid systems rollout. Specifically:

- ***Sumatra*** (population: 45 million). The southern Sumatra Wilayahs of PLN together are akin to the "the next Java-Bali," but with a scaled-down interconnected system
- ***Sulawesi, Kalimantan*** (population: 28 million). Medium-small-scale grid network

35. Financially, PLN operations on Java-Bali are marginally break-even under the allowable tariff level. Compounding this situation, PLN's cost of service outside Java-Bali typically is 20%–30% higher in the more densely populated areas of Sumatra, Kalimantan, and Sulawesi; and between 50%–200% higher in NTB (Nusa Tengara and Maluku) and Papua. As a consequence, PLN is not in a financial position to significantly and rapidly expand access outside Java-Bali. Moreover, from a fiscal and public finance standpoint, PSO subsidy transfers from the Ministry of Finance have risen sharply over the years. In 2009 the transfers reached an all-time high of approximately US$6 billion (for recurrent and investment costs incremental to revenues recoverable by PLN via the uniform national tariff set by GOI). The transfers were estimated in the 2010 budget at US$5 billion, contingent on a realized oil price of $60 per barrel.

solutions especially in Southern Sulawesi and the areas on Kalimantan that have concentrations of economic and demographic density.

In the regions and areas broadly identified above, small-scale decentralized electrification solutions also are worth pursuing at the margins to complement the expansion of the grid system.

- ***Eastern Indonesian Islands: Nusa Tenggara Timur Province (NTT) and Papua.*** The Eastern Islands of Indonesia consist of NTT (population 9 million) and Papua (24 million). These islands have low economic density and comprise primarily widely spread and scattered small settlements. For these islands, a mix of technically feasible and economically viable service delivery modes can be applied. Small grid systems (mostly diesel) are appropriate for major towns and surroundings; otherwise, whatever grids are available: hydro, geothermal, and grid-connected solar PV systems. For the remaining populations living outside grid-accessible areas, portable solar products for lighting, powering cell phones, and powering small radio batteries offer an immediate life-changing option. SHS service delivery programs also can be an effective option. They can be structured not as programs with a focus on technology but as service delivery programs with integrated regular maintenance akin to a utility service. Moreover, they should be targeted on a priority basis to entities that deliver social and administrative services.

Refocus the PSO subsidy financing mechanism to catalyze electrification programs outside Java-Bali within a programmatic framework. The PSO mechanism is a key to effectively address the interlocking issues of maintaining the policy of uniform national tariffs, and to effectively target and subsidize the higher cost of service outside Java-Bali. Essentially, these three goals can be achieved by redesigning the PSO subsidy from PLN's customer categories to PLN regions/Wilayahs. This redesign implies a shift of cross-subsidies across customer classes nationwide to stay within each particular region rather than among regions within customer classes.

A second and key building block, and an essential complement to implement the above strategy, is to advance the formation of standalone business units in PLN Wilayahs. Properly designed and functioning decentralized business units of PLN will enable a verifiable basis on which to establish the differentiation of cost structures across regions outside Java-Bali. Such decentralized units will enable establishing the cost-accounting links to the redesigned PSO mechanism. If designed well and implemented accordingly, such a strategy can yield a "triple win" in meeting the access targets outside Java-Bali consistent with the government's National Electricity Plan timeline. These targets are better poverty targeting, better access targeting, and better green energy targeting.

Setting a few good examples. The strategy outlined above can be tested in a few promising Wilayahs that span the range of cost variation outside Java-Bali. Another potential consideration concerns supportive regional governor(s) who may be willing to champion aggressive implementation backed by financing from provincial/local administration funds. These funds would complement a better targeted PSO subsidy allocation adequate to cover the true margin between PLN's revenue in that region and the cost structure to expand access. Evidence shows that Southern Sumatra and South Sulawesi are strong candidates. It also would be important to apply the redesigned PSO mechanism and strategy in two islands in the Eastern Indonesia region, as these islands are defined by different cost structures and a much higher share of decentralized and small local-grid supply.

A1.3 Lao People's Democratic Republic (Lao PDR)

Over 15 years (1995–2009), electricity access in Lao PDR has more than quadrupled, from approximately 15 percent in 1995 to 70 percent in 2010. Of the latter amount, approximately 2 percent comes from the off-grid program. This expanded electricity access has resulted in over 700,000 HH grid connections by 2010, up from approximately 120,000 HH connected in 1995. The Lao program has maintained a faster pace of implementation than have most other countries. Lao PDR has managed to achieve this result at a relatively low level of GDP per capita, comparable to the experiences of Vietnam and China.

Population *(2008) (mil)*		6.2
Rural population *(% total population, 2008)*		69
Population density *(people/sq km)*		27
Land area *(sq km)*		230,800
GNI per capita *(Atlas Method: Current US$, 2008)*		750
Access to modern cooking fuels		2.6
Urban *(% HH, 2008)*		8
Rural *(% HH, 2008)*		0
Electricity access, national *(% HH, 2009)*		70
Urban *(% HH, 2008)*		84
Rural *(% HH, 2008)*		42
No. of people w/o access to electricity *(2009) (mil)*		1.9
Population served by off-grid sources (minigrids and HH systems) *(%)*		<4
Electricity access target and year *(% HH)*		90; 2020
Electric power consumption *(kWh per capita, 2007)*		243
Installed capacity *(MW 2008)*		723
Thermal		673
Hydro		50
Other renewable		—
Electricity net generation *(bil kWh)*		3.98
Distribution losses *(% net generation)*		7
CO_2 emissions *(M/T per capita, 2007)*		0.3
Indicative residential electricity tariffs for rural consumers *(2011)*	0–25 kWh: 26–150 kWh: Above 150 kWh:	269 lak (USc3.36) 773 lak (USc9.66) 320 lak (USc4.00) (*Source:* www.edl-laos.com)
Key institutions for electricity sector		Ministry for Energy and Mines Electricité du Laos (EDL)

Sources: World Bank 2010e; www.eia.doe.gov; IEA 2010; WHO and UNDP 2009.

IBRD 38384

FEBRUARY 2011

Lao PDR is located in Southeastern Asia, northeast of Thailand and west of Vietnam. The topography of Lao PDR is largely mountainous with elevations above 500 meters, typically characterized by steep terrain and narrow river valleys. This mountainous landscape extends across most of the north of the country, except for the plain of Vientiane and the Plain of Jars in Xiangkhoang Province. The southern "panhandle" of the country contains large level areas in Savannakhét and Champasak provinces. Much of Khammouan Province and the eastern part of all of the southern provinces are mountainous.

Table A1.3.1 Lao PDR: Scenario Analysis for Universal Electricity Access by 2030

Period	Total investment needs (US$ mil)		Incremental needs for universal access (US$ mil)			
	Business-as-Usual scenario	Universal Access scenario	Total	Urban grid	Rural grid	Off-grid
2011–20	470	491	21	—	19	2
2021–30	223	196	—	—	—	—
2011–30	**694**	**687**	—	—	—	—
		Annual requirements:	—	—	—	—

Note: No. of HH without electricity in Business-as-Usual scenario by 2030: <1%.

CHALLENGES

- Grid rollout: Approaching economic limits
- Off-grid program: High expectations; weak institutional and financial framework

STRATEGIES

- Address "last-mile" challenges of grid extension through systematic planning and greater use of proven low-cost solutions

- Off-grid electrification program: Gearing up for acceleration, scale-up, and reach.
- Approach solar home system (SHS) electrification as a program to provide and maintain electricity service, not as an exercise in the procurement and installation of SHS equipment
- Conduct targeted campaigns for SHS electrification based on specific connection targets for delineated areas within each province
- Use EDL as a technically competent entity to implement off-grid programs

Challenges

Lao PDR is on the threshold of graduating from Low Income status. The country's power sector, anchored by the national utility Electricité du Laos (EDL), has been a key partner and enabler of the nation's development. The overwhelming majority of electric connections to date—approximately 68 percent of the 70 percent national coverage—have been implemented by EDL through effectively extending the grid. This extensive use of the grid was made possible through a least-cost expansion plan, which included a simple but rigorous prioritization and village-screening process and deployment of many cost-cutting technical innovations.

As the grid spreads into less accessible and less densely populated areas, the current approach to electrification is becoming very expensive. Looking ahead, new demands and expectations from the sector are posing challenges of greater scale and scope than were encountered earlier. To keep up the current rate of progress in electrification, it will not suffice merely to do more of the same on both the grid and off-grid electrification fronts.

Grid expansion is approaching its economic limits. In Lao PDR, an increasing share of the population without electricity access lives in scattered communities in ever more remote villages, many of which are hard to reach, including mountainous areas. As a result, the average cost of grid extension has nearly doubled from US$450–550/HH 7 years ago, when national access levels were approximately 50 percent, to approximately US$900 today. However, the cost per connection at the fringes of the EDL grid is likely to be 33 percent–50 percent more than the current average and will continue to rise rapidly as the grid is extended.

Off-grid connections program: High expectations and big challenges. The government has set a national target of 90 percent electricity access by 2020. Given the economic limits of grid expansion, off-grid electrification will need to increase 3- to 4-fold over the next decade. This increase translates to an annual off-grid connection rate of 8,000–10,000 connections every year over the medium term. This target is much higher than the current annual rate of approximately 2,000 systems under the SHS program being implemented by the Department of Electricity (DOE).[36] Moreover, the current SHS program framework is already challenged by the rising costs of extending delivery chains beyond the close-in rural areas outside the grid's footprint. The program also is encumbered by an inadequate design and incentive framework that covers multiple agents across an overly extended supply chain. The apex agent lacks the capacity to manage the overall project day to day. Private sector/community-based schemes for mini/micro grids have been planned but are still nascent.

Strategies

Stretching the limits of grid extension: Confronting the "last-mile" challenges. EDL recognizes that the next stage of grid network extensions increasingly will be challenged by population settlement patterns that are less nucleated and require longer line distance-per-kilovolt-Ampere (kVA) connections than typically encountered so far. There is a pressing economic case for systematic planning and widespread deployment of the lowest cost network designs and reticulation that are cost

36. Ministry of Energy and Mines (MEM).

effective for the next stage of grid electrification. In developing its strategic approach to farther grid extension, EDL can benefit from the experience of other nations' electrification programs that are recognized as good practice. These have successfully dealt with "last-mile" access challenges while being mindful of cost-effectiveness and financial viability. Among the proven and promising technical options for grid extensions that have withstood the test of time are conventional single-phase primary network designs and, where appropriate, single-wire earth-return (SWER) spur lines.

Specific cases that are relevant to the Lao PDR context include Brazil, Peru, South Africa, and Tunisia.[37] In all of these countries, single-phase primary networks are a prominent feature of their "last-mile" network design, and helped achieve cost savings of 30 percent–50 percent compared to conventional network design. Furthermore, South Africa and Tunisia in particular have effectively integrated the use of SWER spur lines with single-phase and three-phase medium voltage (MV) primary networks. Together, these networks stretch the reach of the grid coverage to the maximum extent, in some instances reaching loads as remote as 100 km from the MV network.

Off-grid electrification program: Gearing up for acceleration, scale-up, and reach. If it is to succeed at the formidable challenge of quadrupling its present rate of implementing new off-grid connections to meet its overall access target, Lao PDR needs to substantially refocus the strategy and design of its off-grid electrification program. A key first step is to reassess its current pace of implementing off-grid connections. It is important to establish realistic and achievable targets to expand off-grid connections in the context of the overall national access targets. This determination should take into account the relative costs per connection of various technical options, and—critically—the proven implementation capabilities and speed of grid rollout compared to the pace at which primary off-grid agents can be mobilized for the task. This approach clearly goes beyond a narrow, simplistic comparison of the economic cost of new connections at the margins between grid and off-grid options. Finally, a results-driven off-grid electrification strategy and implementation plan that can be counted on to deliver the national targets on time need to be developed. In this context, the following three steps, derived from the experience of better performing off-grid designs in other countries, are proposed for Lao PDR's off-grid implementation program.

- ***Ongoing assessment of the off-grid "space" and grid interface.*** If EDL were to embrace a systematic and widespread deployment of lower cost single-phase primary network design and reticulation and maximize the use of SWER line spurs where appropriate, the nation could greatly expand the reach and coverage of its grid. This type of deployment clearly would require updating the off-grid program targets each year. To continually make this assessment, it is crucial to have a nationwide spatial grid expansion plan that is updated continuously and reflects EDL's grid extension strategy as it is implemented. A planning and rapid appraisal platform especially relevant for developing such a plan is available from the recent generation of GIS-based spatial planning models. In contrast to the traditional (and cumbersome) Master Plan analysis approach, analysis that is based on a GIS-based appraisal platform provides a quantitative and representative spatial reference frame for (a) tracking and updating

37. For example, Brazil has achieved near-universal access to electricity, with 97% or more coverage provided by grid-based electricity. The last 4% of coverage was obtained through a conscious strategy of grid extensions, using single-phase network design to the maximum extent feasible. When unit costs breached the $4,000 per connection zone, isolated diesel mini-grids were deployed. Now that near-universal access has been achieved, an ongoing program converts as many of the diesel isolated grids as possible to gas or other cost-competitive fuels if available locally.

ongoing actual progress and (b) more sharply delineating the evolving off-grid program space in relation to the grid. (See box 2.5 for an illustration of the GIS-based spatial planning technique.)

- ***Solar home system (SHS) electrification program: Targeted campaigns with connection targets, staged by delineated areas within each province.*** In seeking to expand its SHS electrification program, Lao PDR can learn from Peru's main off-grid electrification program. It offers particularly relevant lessons and design insights for developing a program of the required scale, speed, and enhanced spatial reach and coverage.

  Peru's national electrification coverage is approximately 80 percent–95 percent in urban areas, and approximately 35 percent in rural areas—mostly from the grid. The limits of grid coverage from the ongoing electrification are projected to be 90 percent. The reason is that large parts of the country consist of highly challenging mountainous terrain with spatially dispersed small settlements akin to the northern Lao PDR region. For these settlements, in parallel to its grid rollout program, Peru is orchestrating a well-planned, targeted, and structured off-grid electrification program. Its main driver is an SHS-based electrification service targeted to reach approximately 300,000 HH and staged in a programmatic manner.

  Under Peru's approach, qualified service providers undertake SHS electrification of targeted areas within an established technical and regulatory framework that covers equipment and service standards, and modalities for financing and subsidy.

  The objective is not a program paradigm designed and managed to procure and install SHS equipment. The objective and guiding philosophy of this program is to provide and thereafter maintain electricity service. In keeping with this approach, customers pay an equivalent monthly tariff for an electricity utility service type of arrangement.

  Bids from qualified service providers—private sector, NGOs, and other qualified agents—are evaluated on a competitive basis for each package. To date, a notable feature of Peru's off-grid program design and experience is that the SHS electrification bid packages also are open to the regional electric utilities. Indeed, in several instances, utilities have won the bid. The utilities then undertake all of the key functions including overall project management of the delivery and installation of the systems, and, significantly, manage the ongoing provision of the electricity service.[38]

- ***Utilities can be effective implementation agents for off-grid programs.*** In addition to Peru, Argentina and Brazil offer instructive examples for the Lao PDR off-grid program. In many countries, electricity utilities have been effectively mobilized, together with the private sector, to implement one or more off-grid program components. Brazil's off-grid program covers over 3,000 isolated diesel minigrid systems that are embedded in the respective service areas of the distribution utility serving that area. A nationally operated cross-subsidy mechanism funds the Universal Access fund, which compensates each utility for 90

38. In Peru, it is common to find instances in which the regional electric utility proximate to the bid area is the winning bidder. The utility is already familiar with the area demographics and other characteristics. Moreover, it has the capacity to undertake effectively project management and, later, maintenance and commercial functions. The reason for the utilities' success is that an integral part of their core business model for grid-based services has been outsourcing a broad range of technical and commercial activities to small private enterprises. The regulatory and subsidy arrangements for the solar PV connections allow for a full cost recovery tariff (as for grid connections) and are financially structured to enable the most efficient agents to make a profit. In many instances—although not in every case—the regional local electric utility has both a competitive edge and an interest.

percent of the allowable investment cost in addition to any operating losses. Tariffs are more or less uniform across the nation. A significant component of Argentina's last-mile electrification strategy is the staged implementation of a SHS electrification program with ongoing maintenance service charged on a monthly billing basis, akin to a normal grid customer's monthly bill. In Argentina, the regional electric utility companies have set up subsidiaries whose core business is to manage and operate the off-grid SHS electrification service program.

In short, established national and regional utilities—public or private—with proven track records represent valuable national assets that could be deployed to implement one or more off-grid program components. These technical and management capacities, and the professionalism and results-oriented culture, of these utilities can be harnessed under a supportive regulatory, policy, and incentive framework for effective off-grid program delivery.

From these three steps, it follows that there is a strong case for establishing an EDL subsidiary whose core business would be solely off-grid project design, management, and operations. This subsidiary would undertake this program on a fully costed fee basis charged to the Department of Energy. Subsidy funds for a scaled-up off-grid program can be sourced from a properly designed and capitalized Rural Electrification Fund and disbursement mechanism. Such a subsidiary would be open to compete with other qualified agents for a range of functions and services. These could include project management and oversight for scaled-up SHS program design, installations, and services and maintenance thereafter; as well as minigrid design and construction oversight.

A1.4 Mongolia

Approximately 90 percent of Mongolia's population has access to electricity, but with wide disparities between urban and rural areas. Nearly 99 percent of urban residents, including Ulanbaatar and the *aimag* (province centers), have access to electricity; compared to 80 percent of *soum* (district) center residents and to only 25 percent of the herder population of approximately 850,000 (approximately 30 percent of the country's population). In the soum centers, electricity is provided through isolated grids based on diesel systems. The electricity is of significantly lower

Population *(2008) (mil)*	2.6
Rural population *(% total population, 2008)*	43
Population density *(people/sq km)*	2
Land area *(sq km)*	1,560,000
GNI per capita *(Atlas Method: Current US$, 2008)*	1,679
Access to modern cooking fuels	23
Urban *(% HH, 2008)*	31
Rural *(% HH, 2008)*	2
Electricity access, national *(% HH, 2009)*	90
Urban *(% HH, 2008)*	99
Rural *(% HH, 2008)*	75
No. of people w/o access to electricity *(2009) (mil)*	0.63
Population served by off-grid sources (minigrids and HH systems) *(%)*	20
Electricity access target and year *(% HH)*	100; 2020
Electric power consumption *(kWh per capita, 2007)*	1369
Installed capacity *(MW 2008)*	832
Thermal	832
Hydro	—
Other renewable	—
Electricity net generation *(bil kWh)*	3.9
Distribution losses *(% net generation)*	11
CO_2 emissions *(M/T per capita, 2007)*	4.0
Indicative residential electricity tariffs for rural consumers *(2011)*	US 2.9c/kWh in all aimags and soums if under 50kWh/month (*Source:* Energy Regulatory Authority Annual Report 2009)
Key institutions for electricity sector	Ministry of Mineral Resources and Energy (MMRE) Electricity Authority (EA) Electricity Regulatory Authority

Sources: World Bank 2010e; www.eia.doe.gov; IEA 2010; WHO and UNDP 2009.

IBRD 38385

FEBRUARY 2011

Mongolia is a landlocked country whose terrain comprises mountains and rolling plateaus, with a high degree of relief. Overall, the land slopes from the high Altay Mountains of the west and the north to plains and depressions in the east and the south. The country has an average elevation of 1,580 meters. The landscape includes one of Asia's largest freshwater lakes (Lake Khövsgöl), many salt lakes, marshes, sand dunes, rolling grasslands, alpine forests, and permanent mountain glaciers. Northern and western Mongolia are seismically active zones, with frequent earthquakes and many hot springs and extinct volcanoes.

Table A1.4.1 Mongolia: Scenario Analysis for Universal Electricity Access by 2030

	Total investment needs (US$ mil)		Incremental needs for universal access (US$ mil)			
Period	**Business-as-Usual scenario**	**Universal Access scenario**	**Total**	**Urban grid**	**Rural grid**	**Off-grid**
2011–20	88	88	—	—	—	—
2021–30	38	38	—	—	—	—
2011–30	**126**	**126**	—	—	—	—
		Annual requirements:	—	—	—	—

Note: No. of HH without electricity in Business-as-Usual scenario by 2030: <1%.

duration, quality, and reliability compared to Ulanbaatar and the aimag centers. With limited or no access to electricity, rural (mostly remote soum centers, and herder communities) living standards are constrained by low agricultural and livestock productivity and very few opportunities for nonfarm employment or other value-adding economic activities.

Sector Structure

Mongolia's Ministry of Mineral Resources and Energy (MMRE) has overall responsibility for the power sector. The Energy Authority (EA), which reports to MMRE, is charged with policy-making and project implementation. The Energy Regulatory Authority oversees licensing, tariff-setting, and promotion of competition. In 2001, to improve the efficiency of the sector, the power sector was unbundled into 18 generation, transmission, and distribution companies. However, these reforms have yet to yield the desired results.

Consumer Groups

The power sector in Mongolia is segmented into three consumer groups: urban, soum (district), and herders.

- ***Urban.*** The urban consumer group is the largest and includes all grid-connected cities and towns. This group is served by three grids: Central, which supplies Ulaanbaatar; Eastern; and Western Energy Systems. These three supply mostly coal-based power and account for approximately 97 percent of the electricity sales in Mongolia.
- ***Soums.*** Of the 314 soum centers in the country, only 117 are connected to the national or regional grids. The remaining 197 soum centers depend on small diesel generators and microgrids, which are owned and operated by local governments. These small systems provide at most 4–6 hours a day of electricity to the soum residents. Several off-grid soum centers have been provided with new "free" diesel generators through bilateral assistance, but, otherwise, local governments have been left to fend for themselves in operating and maintaining these generators, and financing operating costs.

  To increase rural coverage and improve the cost-effectiveness of the isolated grids in remote soum centers, the government's National Renewable Energy Program has set ambitious goals for broad-based renewable energy development. These goals are to increase the share of renewable energy technologies in total energy supply from 0.9 percent in 2005 to 3.0 percent–5.0 percent by 2010; and to 20.0 percent–25.0 percent by 2020.
- ***Herders.*** This group consists of nomadic herders. Household-based power supply systems are the only option for this market. At present, approximately 60,000 independent solar PV systems are being used by herders to operate lights, radios, televisions, and satellite dishes. Under the "100,000 Ger" program, the government has financed, procured, sold, and installed over 40,000 SHS.[39] The government is on track to reach 92,500 in the next phase, operating through a combination of the soum administration network, private dealers, and sales and service centers (SSCs). An additional 60,000 SHS are expected to be delivered through the WB-financed Renewable Energy for Rural Access Project (REAP).

Rural Targets

The government has assigned off-grid soum and HH electrification an ambitious target of 150,000 HH by 2020. This target envisages that up to 20 percent of the rural population located in isolated areas can be powered with indigenous energy resources, including solar, picohydro generator units, biomass, and wind.

Use of Off-Grid Renewable Systems

The government plans to expand the use of off-grid renewable systems while preserving service quality and affordability. The strategy is to expand electrification while improving service quality and affordability through (1) increasing electricity access through SHSs and small wind turbine systems (WTSs) and (2) developing renewable or renewable-diesel hybrid systems among off-grid soum centers. The government also aims to improve the financial viability of the soum center electricity service by making it more commercially oriented and encouraging private sector and community participation in providing electricity service.

CHALLENGES

- Low capacity in the Energy Authority (EA) for integrated rural access planning

39. A ger is a felt-lined tent home used by Mongolian nomads.

- Low financial and operating viability of most soum center minigrids
- Lack of effective mechanism for SHS/wind system rollout

STRATEGIES

- Improve planning capacity through targeted external technical assistance and expertise
- Improve financial and operating viability for soum grids
- Develop a service-oriented approach to SHS and wind turbine systems
- Secure concessional finance for wind-diesel and wind-solar-diesel hybrid power stations by mobilizing a common platform of donors and leveraging climate change funds

CHALLENGES

Low capacity for planning in the Energy Authority (EA) in the overall context of integrated rural planning. Soum-based rural electrification projects are constrained by weak rural electrification planning capacity in the Energy Authority. More generally, in the absence of integrated rural development planning, income-generating electricity uses have not increased enough to make rural power supply financially viable.

Lack of financial viability and operating capacity of soum center minigrids. Soum center minigrids display generation inefficiencies and high distribution losses. In addition, generation costs are high due to diesel prices. Tariffs do cover the operating costs. There are no clear tariff-setting mechanisms or rate classes among customers. For the most part, electricity consumption is not metered, except for large public institutional facilities. It is estimated that rural HH consume 80 percent of the total electricity while providing approximately only 33 percent of the total revenues. In other words, rural HH effectively are being cross-subsidized by the rural public institutions. Soum centers must pay for diesel in cash. When the cash cost of operation goes up due to increases in diesel prices, and if the customers collectively cannot pay the additional costs, the hours of service are reduced to cut overall fuel costs.

The government plans to connect most off-grid soum centers with national or regional grids in the next 10 years or so. However, approximately 70 off-grid soum centers are considered uneconomical for grid connection due to their remoteness and will need financially sustainable solutions to improve the reliability of their services.

Lack of effective mechanism for SHS/wind system rollout. Multiple factors have inhibited the expansion of the herders' SHS market:

- The incomes of the target group are low. Approximately 40 percent of the estimated 170,000 herder HH have annual cash incomes below US$450 (2006) and cannot afford the capital costs of solar home systems (SHS), which today normally would exceed $500.
- It is difficult to develop an effective supply chain to support a decentralized market for a small and mobile customer base spread over a vast area.
- Basic quality and services standards are lacking for SHS.
- Even though there is a growing microfinance market, because of the unpredictable market and regulations and the perception of high risks in investing in the supply chains, the private sector does not participate in the SHS market.
- Private sector capacity for scaling up implementation of rural electrification through retail systems for private HH and public institutions or for soum center grid electrification is limited.
- Financing, whether grant or concessional debt, for RE is insufficiently available.

Approximately 40,000 HH solar PV and wind power systems and 3,000 wind turbine systems (most of which were donated by the governments of China and Japan) have been sold in Mongolia, sometimes with heavy and poorly targeted subsidies. Some public investments have been made in PV, wind, or hybrid systems for institutional use (schools, hospitals).

Strategies

Improve EA's technical and planning capacity. External TA is needed to develop the capacity for technical operation and management in the relevant government departments, the Energy Authority, and the independent regulator. Without this capacity, it will be difficult to meet the government's access targets.

Improve financial and operating viability for soum grids. Improving the technical and financial performance of the soum microgrids is the first step toward universal HH coverage within the soum center areas. To improve their performances, Mongolia needs to (1) adopt a rational tariff and subsidy policy based on clear performance benchmarks; (2) reorient soum center electricity service to be more commercial; and (3) actively pursue private sector and community participation in the provision of electricity service. Additional actions to strengthen the viability of soum microgrids would be to (4) invest in network rehabilitation and (5) develop generation options with lower operating costs, including full or hybrid renewable energy power generation.

Develop a service-based approach for SHS and WTS. One option is a shift to provide off-grid electrification as a *service,* not as a program paradigm designed and managed to procure and install SHS or other off-grid equipment. In keeping with this approach, for being provided with reliable service from off-grids sources, customers would pay a monthly tariff as is done for grid electricity. The institution charged with providing the service must have adequate capacity for this task. Such a service-oriented approach would need to be anchored by an institution with adequate capacity and familiarity with the consumer base.

Secure concessional finance for renewable energy and hybrid power stations by mobilizing a common platform of donors and leveraging climate change funds. Expanding the base of renewable energy will need substantial investments that will require external financial and technical assistance. Mongolia should develop a phased spatial electrification rollout plan to cover the unconnected HH. Such a plan could be the basis for obtaining external programmatic assistance. The plan also could be used to leverage climate change funds that support renewable energy.

A1.5 The Philippines

Population *(2008) (mil)*	90.4
Rural population *(% total population, 2008)*	34
Population density *(people/sq km)*	303
Land area *(sq km)*	300,000
GNI per capita *(Atlas Method: Current US$, 2008)*	1,700
Access to modern cooking fuels	49
Urban *(% HH, 2008)*	77
Rural *(% HH, 2008)*	29
Electricity access, national *(% HH, 2009)*	84
Urban *(% HH, 2008)*	97
Rural *(% HH, 2008)*	65
No. of people w/o access to electricity *(2009) (mil)*	9.5
Population served by off-grid sources (minigrids and HH systems) *(%)*	<2
Electricity access target and year *(% HH)*	90; 2017
Electric power consumption *(kWh per capita, 2007)*	586
Installed capacity *(MW 2008)*	15,680
Thermal	10,397
Hydro	3,291
Other renewable	1,992
Electricity net generation *(bil kWh)*	59.2
Distribution losses *(% net generation)*	13
CO_2 emissions *(M/T per capita, 2007)*	0.8
Indicative residential electricity tariffs for rural consumers *(2011)*	First 20 kWh free for lifeline; next 30 kWh @ 50% of general tariffs (*Source:* Meralco)
Key institutions for electricity sector	Department of Energy National Electrification Administration (NEA) National Power Corporation (NPC) Rural electric cooperatives (RECs) Small Power Utilities Group (SPUG), division of NPC Energy Regulatory Board Transmission Corporation

Sources: World Bank 2010e; www.eia.doe.gov; IEA 2010; WHO and UNDP 2009.

The Government of the Philippines (GOP) plans to increase the share of electrified households[40] in the country from the present level of 84 percent to 90 percent by 2017. For rural areas, the Philippines Energy Plan (2009–30) envisions reaching 90 percent household electrification,

40. Targeting connections is relatively new for the country, as the main goal of electrification policy had been the interim target of achieving full *barangay* (local district) electrification. The DOE considers the goal of full barangay electrification to have been reached this year (2010). In practice, a barangay is considered electrified if a very basic level of service has been established (for example, the barangay district office has a connection and/or a street light has been installed).

The Philippines is an archipelago comprising 7,107 islands. The 11 largest islands contain 94 percent of the total land area led by Luzon at 105,000 sq km, followed by Mindanao at 95,000 sq km. The islands are divided into three groups: Luzon, Visayas, and Mindanao. The Luzon islands include Luzon Island itself, Palawan, Mindoro, Marinduque, Masbate, and Batanes. The Visayas in the central Philippines contain Panay, Negros, Cebu, Bohol, Leyte, and Samar. The Mindanao islands comprise Mindanao itself plus the Sulu Archipelago.

Table A1.5.1 Philippines: Scenario Analysis for Universal Electricity Access by 2030

	Total investment needs (US$ mil)		Incremental needs for universal access (US$ mil)			
Period	**Business-as-Usual scenario**	**Universal Access scenario**	**Total**	**Urban grid**	**Rural grid**	**Off-grid**
2011–20	3,605	3,605	—	—	—	—
2021–30	2,340	3,408	1,068	—	871	198
2011–30	**5,946**	**7,014**	**1,068**	—	**871**	**198**
		Annual requirements:	53	—	44	10

Note: No. of HH without electricity in Business-as-Usual scenario by 2030: 1 million (<4% of the total population).

from the present estimated level of 65 percent. To reach these targets, electricity will have to be provided to 3.4 million households over 2000–17. Generation capacity is presently a constraint. To respond, under the Missionary Electrification Development Plan (2009–18), the GOP intends to expand electricity generation capacity in approximately 200 nonelectrified areas through subsidies and private sector participation programs. Over 2009–17, nearly 1 million households will have be served by solar home systems (SHS) because it is not possible to connect these households at reasonable cost to either the major regional grids or any of the isolated systems in remote locations throughout the archipelagic nation.

Background

Institutions. Several agencies are involved in providing power supply to rural and remote locations. These agencies include the Small Power Utilities Group (SPUG) of the National Power Corporation,[41] 119 rural electric cooperatives (RECs), local government units (LGUs), private distribution companies, and "qualified third parties" (QTPs).[42] SPUG operates in 78 islands and an additional 8 isolated areas on major islands. It has over 300 generators totaling approximately 129 megawatts, almost all of which are diesel fired. Most of the buyers of SPUG's generated power are RECs and private QTPs. LGUs no longer are major buyers of SPUG power.

Over the years, the government has undertaken several institutional reforms to meet increasing demand and improve performance, particularly through the 2001 Electric Power Industry Reform Act (EPIRA). Several LGUs are small and loss making. If potentially viable, their operations are sold off. If not, the assets typically are transferred to SPUG or the RECs. A handful of QTP operations still exist in the country, although not nearly so many as the government had hoped for. Combined, the state agencies, RECs, and QTPs have considerable capacity and experience to provide electricity services.

Rural electric cooperatives. In the rural Philippines, power distribution is under the purview of 119 rural electric cooperatives (RECs), although RECs do not serve exclusively rural areas. RECs are supervised by NEA, which traditionally has provided technical support and financial support to them. However, since 2001, EPIRA has been expected to prepare the RECs for the implementation of various industry reforms, including wholesale electricity market trading (WESM) and retail competition. RECs' customer bases vary widely from 1,000 to over 130,000 customers each.

RECs' financial performance has been uneven due to low cost recovery from socially acceptable tariffs, poor financial management, decline in bill collection rates, and increasing political interference. Regardless of the conditions in its service territory, each REC is required to meet the same financial efficiency standards. This requirement imposes financial strains on RECs that must operate under unfavorable conditions. In fact, in some cases, regulatory pressure on rates can reduce efficiency by preventing RECs from spending adequately for maintenance and renewals. It is estimated that, given current tariffs, 33 percent of RECs are not financially viable; and many do not meet the technical and financial requirements necessary to recover costs. Options to improve the performance of RECs include reorganizing and/or merging RECs into viable units, or converting them into joint stock companies. However, these approaches can be difficult to implement.

Improving governance and management is key to improve the performance of RECs, which are independent entities that have the freedom to make decisions on most issues. For a well-run REC, access to capital should not be a constraint. Potential sources include NEA, the Development Bank of the Philippines (DBP), LGU Guarantee Corporation (LGUGC), and the Rural Electrification Financing Corporation (REFC). Nevertheless, the reality is that several RECs need to improve their operations, finances, and governance in some combination to become attractive even to public sector lenders such as NEA.

The ending of the government role as supplier of last resort through the National Power Company/Power Sector Assets and Liabilities

41. SPUG is a division of NPC responsible for generating electricity for sale to electric cooperatives (ECs) serving areas that, due to their remoteness, are unable to access a grid connection.

42. A QTP is a private company or organization that generates and distributes, or distributes only, services in a franchise area that either has been ceded by the local electric cooperative, or is an unserved territory not associated with an EC, in which electricity service is being established under the GOP's Missionary Electrification Development Program. ("Missionary" means essentially "unserved.")

Management Corporation (NPC/PSALM) presents an opportunity to improve accountability and performance in RECs. A utility that does not pay for electricity supply risks being cut off, and, perhaps more important from the viewpoint of an individual REC, risks having NEA exercise its step-in rights to replace existing management and dissolve the board. This prospect, which has materialized in some cases, has significantly improved the RECs' performance.

CHALLENGES

- Address large financing gap for operating costs and new investments in rural areas
- Improve REC performance
- Promote private sector role in rural electrification.

STRATEGIES

- Develop well-costed and time-bound spatial electrification plan as basis for channeling programmatic assistance
- Explore opportunities for output-based aid
- Move toward tailored standards for distribution losses and reinvestment allowance for RECs
- Encourage "service" approach for off-grid renewable applications

Challenges

Addressing large financing gap for operating costs and new investments in rural areas. This mismatch between policy objectives and the level of allocated resources is a critical barrier to expanding rural electrification in the Philippines. Under almost any circumstance, though, there is likely to be a significant funding gap that cannot be filled without the government's committing additional funds.

DOE's electrification plan calls for significantly expanding SPUG activities to over 70 new areas. Doing so would require an additional 54 MW of new generating capacity. However, SPUG's poor financial condition makes this expansion extremely difficult. Missionary electrification programs are expected to be financed through the "universal charge for missionary electrification" (UC-ME), which is added to the bills of all customers on the integrated networks. The main use of UC-ME revenue is to cover SPUG's operating losses. However, added to the fact that the level of actual UC-ME revenue and the timeliness with which SPUG receives it are not adequate, there simply is not enough money to support electrification expansion programs.

SPUG's annual subsidy requirement is estimated by DOE at 18 billion pesos (US$400 million). Most of this amount is intended to pay for oil-fired generation and reflects the difference between the cost of generation, which is at least 20 pesos (45 cents) per kWh, and the socially acceptable charge for service of approximately 8 pesos (18 cents) per kWh.

Improving REC performance. Despite overall improvements in recent years, REC performance continues to be constrained by governance issues, inadequate management, and poor technical performance.

Promoting private sector role in rural electrification. The government's policy provides for a substantial private sector role. However, in practice, the private sector has a limited appetite for rural electrification, whether as generators, distributors, or SHS providers. Green-site mini/microgrids with private sector/community-based schemes as a service delivery model operate outside mainstream frameworks and are still few in number.

Part of the reason for the reluctance is that GOP programs are limited, often by the amount of money that can be attracted from specific donors such as GEF, the World Bank, and ADB. However, even for limited operations, private and public operators alike must deal with difficult and expensive supply chains; and service territories with difficult geography, dispersed and poor households, and limited economic growth prospects. Many remote regions also have a law and order problem. Finally, even if a company wanted to be in the business of providing energy services to remote areas, it is very difficult to find qualified people willing to do this kind of work.

Strategies

Develop a well-costed and time-bound spatial electrification plan as the basis for channeling programmatic assistance. External assistance will be required on a significant scale to supplement government funds for making necessary investments in generation and off-grid technologies and for distribution of assets. A pragmatic approach would be to prepare a spatial and time-sequenced plan that would (1) systematically identify the communities and clusters that need to be served—by both minigrid and household systems; (2) estimate costs; (3) identify subsidy requirements and sources; and (4) sequence investments. Such a plan could provide an improved basis for attracting programmatic lending from donors and sources of concessional finance. For the greatest impact, these sources of funds then should be passed on at similar concessionary terms for investments.

Explore opportunities for output-based aid. The Philippines' rapid rate of urbanization suggests that many of the millions of future connections will be made by distribution companies in urban and peri-urban areas. A large proportion of the projected 1 million SHS connections will be very expensive. Output-based aid can be a promising option to (1) subsidize the initial connection and (2) develop a regulated utility model so that service providers can be fairly compensated for ongoing services and held accountable to specific service standards. A well-designed OBA model could attract multilateral and bilateral donor assistance.

For RECs, move toward tailored standards for distribution losses and reinvestment allowance. RECs vary widely in size and in the geographical conditions under which they operate. Better performing RECs serve large and progressive provincial cities, whereas the poor-performing RECs normally serve small islands or those with law and order problems. A regulatory approach of stipulating common standards for distribution losses and reinvestment allowances is ineffective in promoting efficiency On the contrary, regulatory pressure on rates can reduce efficiency by preventing RECs from spending enough on maintenance and renewals. The regulator (ERC) should undertake a detailed examination of costs for each REC. ERC then should create efficiency standards that are specific to each entity against which to judge its performance and to provide a basis for "smart" subsidies. The regulator already has moved toward a performance-linked benchmarking system for the ECs that differentiates among the cost structures of individual ECs.

Encourage "service" approach for off-grid renewable applications. It will be important to leverage the strengths of existing entities to design and implement off-grid investments and services, such as minigrids and SHS systems design, technical oversight, investment rollout, and operation of off-grid services. In contrast to a "dealer" approach, a "service" approach should be tried for off-grid HH applications in which the customer pays a monthly fee for equipment that is maintained by the service provider. Peru has such a system. In Argentina, the regional electricity utility companies have set up subsidiaries whose core business is to manage and operate the off-grid SHS electrification service program. The Philippines has made a start in this regard with a few RECs' attempts to install approximately 2000 SHS.

A1.6 Pacific Island Countries (PICs)

Shared Context: Economic Geography Shaping Development Challenges Far More Than Elsewhere

The 12 Pacific Island Countries (PICs) in the World Bank's East Asia and the Pacific Region include approximately 10 million people (approximately 2.2 million people if Papua New Guinea and Timor-Leste are excluded).[43] Of the 12 PICs, 8 have populations well below 200,000 (table A1.6.1). The per capita GNI in most of these countries is US$4,000. Several are characterized by high population growth rates. Literacy rates exceed 90 percent in most of the countries, excluding PNG, Timor-Leste, and Vanuatu. The Region is relatively aid abundant on a per capita basis, and several of these island economies rely heavily on development cooperation flows. In particular, the North Pacific region states receive some of the highest levels of aid per capita of any countries in the world (up to as much as US$1,000 per capita, or approximately 50 percent of the GNI per capita). However, in many PICs, development results in both service delivery and growth have not matched the aid (table A1.6.1).

The citizens of most Pacific Island nations put up with low access to secure, reliable, and affordable energy. Of the nearly 10 million people living in the Pacific Island Countries, an estimated 8 million do not have access to electricity (0.9 million excluding PNG and Timor-Leste) (table A1.6.2).

As a group, PICs have some unique features. For one, in most instances, even the fortunate few with grid service pay unit prices for electricity that rank among the highest in the world (25–50+ USc/kWh). High unit prices pose a significant affordability barrier and overly restrict household consumption and its derived benefits. High electricity tariffs also put additional pressure on enterprises by raising the already high costs of doing business. Finally, those not connected to the grid incur even higher costs by self-generating with heavily taxed fuels.

The high grid electricity prices are due in part to the dependence on high-cost diesel-based generation and, in some instances, to avoidable operational inefficiencies, particularly high network losses and high unit fuel consumption rates.[44] Many grid utilities also have chronic problems of service unreliability and poor revenue management. Some are substantially dependent on recurrent subsidies from their governments. These subsidies take the form of direct transfers or contingent liabilities arising from financial guarantees provided to utilities to procure oil. Such direct and implicit subsidies often are significant in magnitude, adding to the countries' fiscal and macroeconomic management challenges.

Another special facet of the PICs is the challenge of meeting even the most basic electricity services needs of two large populations. One group lives outside the grid coverage today and for the foreseeable future. The second group is scattered widely across far-flung islands and remote small communities. Combined, these 2 groups comprise 80 percent of the PICs population on a regional basis (41 percent excluding PNG and Timor-Leste) (table A1.6.3). The extent of spatial dispersion of settlements in the PICs contrasts sharply with the situations of most other countries, whose geographic and settlement patterns generally permit 85 percent–95 percent of the population ultimately to be connected to the grid through a least-cost grid rollout plan.

This chapter highlights the key relevant strategic and policy directions worthy of consideration in addressing the interlocked set of challenges to secure, affordable, and widely accessible quality energy services for all PIC citizens. These

---

43. The World Bank's East Asia and the Pacific (EAP) Region comprises 12 Pacific Island Country members: Melanesia (Fiji, Solomon Islands, and Vanuatu), Polynesia (Samoa, Tonga, and Tuvalu), Micronesia (Kiribati and Republic of Marshall Islands, or RMI), Federated States of Micronesia (or FSM, and Palau), Papua-New Guinea (PNG), and Timor-Leste.

44. There is considerable scope for improvements in the customer-side efficiency of electricity use as well.

Table A1.6.1 EAP Region Pacific Island Countries Selected Indicators

Country	Population 2009 ('000s)	Population growth rate, 1990–2008 (%/yr)	GNI per capita 2008 (Atlas Method) (US$)	GNI per capita rank (US$)	Aid per capita 2007 (US$)	Literacy 2008 (%)
Fiji	843	0.8	3,930	112	69	94
Kiribati*	100*	1.6	2,000	143	285	—
Marshall Islands	60	1.4	3,270	124	894	94
Micronesia, Fed. States	111	0.8	2,340	138	1,035	89
Palau	20	1.6	8,650	79	1,108	92
Papua new Guinea	6,599	2.5	2,030	168	50	57
Samoa	179	0.7	2,780	129	204	~100
Solomon islands	510	2.7	1,180	156	501	—
Timor-Leste	1,098	2.2	2,460	136	262	59
Tonga	103	0.5	2,560	133	296	99
Tuvalu	12	—	—	—	—	—
Vanuatu	233*	2.4	2,330	139	251	74

Source: World Development Indicators 2009 database.
Notes: * = Kiribati: World Bank estimates 2010; Vanuatu: Census update December 2010.

suggestions are grounded in established good practice lessons drawn from relevant worldwide experience. These lessons include regional good practices, especially experience with deployment of grid-renewable energy generation systems on the scale of island utilities that can substantially reduce the diesel dependency and lower unit costs of Pacific Island grid systems.

For the nearly 80 percent of the PICs population who live in dispersed settlements outside of grid system footprint areas, of particular significance are the experience and trends in the market for consumer friendly "picosolar" products. These relatively low-cost, user-friendly gadgets provide high-quality modern light-emitting diode (LED) lighting and sufficient power for cell phone and small radio battery charging. Taken together, this "basic electricity" service potential represents an immediate life-changing prospect for off-grid populations (box 3.5).

Remote and "sea locked." The total land area of the PICs (excluding PNG), comprising hundreds of small islands and atolls, is only 89,000 square kilometers (sq km), or approximately one-third the size of New Zealand. This land area is widely scattered across an ocean area that is equivalent to approximately 15 percent of the earth's surface. The physical distances from the world's trade centers make these populations among the most isolated anywhere. For most of these islands, the closest major ports and export markets are Auckland, Sydney, or Tokyo, which are over 3,000 km away on average. The least remote island, Palau, is 1,677 km from Manila (figure A1.6.1). As a consequence, the petroleum supply chains to the PICs are among the most extended and have low economies of scale.

Fragmentation and remoteness within. Significantly, in the majority of the small island nations,

Table A1.6.2 EAP Pacific Island Countries Electricity Grid Access and Cell Phone Coverage, 2009

Country	Electricity access *(%)*	Population without electricity *(000s)*	Teledensity *(cell phones/ 100 pop)*
Fiji	75	211	75*
Kiribati	45*	55	1
Marshall Islands	50	30	1
Micronesia, Fed. States	45	61	25
Palau	95	1	53
Papua New Guinea	under 10	6,137	—
Samoa	95+	9	60*
Solomon islands	Under 20	408	2
Timor-Leste	Under 20	878	7
Tonga	95+	5	60*
Tuvalu	95+	1	—
Vanuatu	27	170	60*
Total *(excl. PNG and Timor-Leste)*	**20** *(59)*	**7,903** *(1,761)*	—

Source: World Development Indicators 2009 database.
Note: * = Entries are for 2009.

less than half the population lives on the main island; and, typically, less than 30 percent lives in cities. PNG's rugged geography means that the majority of people live in highly remote communities with almost no road access to neighboring districts and with very distinct cultural and linguistic practices. For instance, in PNG, the large number of highly fragmented communities comprise over 1,000 languages in PNG, Solomon Islands, and Vanuatu. Another illustration of remoteness within a country is found in Kiribati. Its population of approximately 100,000 is spread over 33 low-lying atolls scattered over 4,000 km of ocean from east to west, and 2,000 km from north to south (a sea footprint equivalent to the land area of the lower 48 states of the United States). The 233,000 inhabitants of Vanuatu are spread over 80 mostly volcanic islands in an 800-km north-south-aligned chain (figure A1.6.1).[45]

Economic isolation. Perhaps more than any other countries, the Pacific Island nations are shaped by their economic geographies. In general the correlation between access to markets and economic growth is strong. Besides their small market sizes, PICs' long distances from main centers of economic activity have hampered their development. The *World Development Report 2009* (WDR) presents several potential market remoteness measures for the Pacific Island Countries. The mean GDP-weighted distance rank for all the islands is 197 of 219 countries. Micronesia is ranked at 176; Polynesia and Melanesia tied at rank 207.[46]

PIC governments are especially challenged to deliver vital services to their far-flung citizens. PICs' geographic constraints not only have played a major role in shaping economic opportunities in the Pacific but also pose special challenges to the provision of basic infrastructure delivery and other priority services. As noted above, populations are widely dispersed and located at great distances from capital cities and economic hubs; transportation (intra- and interisland) as well as communication and information flow links tend to be limited and expensive; and regional and

45. Melanesia is characterized by larger land masses and populations, Polynesia by smaller but relatively compact land masses, and Micronesia by very small and widely dispersed land masses. Melanesia tends to be relatively resource rich; Polynesia generally possesses adequate resources; Micronesia tends to be resource poor.

46. Overall, the PICs are considerably smaller and far more remote from key export markets than are the Caribbean islands. The approximately 40 million people in the Caribbean live across a sea area equivalent to 0.5 percent of the world's surface area and are fewer than 1,000 km from the huge US market. The 2009 WDR estimated the Caribbean island country remoteness index mean to be 100 of 219.

Table A1.6.3 EAP Pacific Island Countries: Physical Geography and Population Settlement Patterns

Country	Pop. *('000s)*	Avg. HH size	Area *(000 sq km)*	Urban pop. *(%)*	Selected physical and spatial geography characteristics
Fiji	843	4.8	18.0	52	320 islands (approx. 100 inhabited). Largest 2 islands comprise approx. 87% of land and 95% of pop. (Viti Levu, Vanua Levu).
Kiribati	100*	6.3	0.8	44	32 widely scattered low coral atolls in 3 groups and 1 raised coral island. They stretch 4200 km east-west, 2000 km north-south. Approx. 48% of pop. resides on islets of S. Tarawa atoll.
Republic of Marshall Islands	60	7.2	0.2	71	Two archipelagic island chains of 29 atolls; (22 inhabited) and 5 raised coral islands (4 inhabited).
Fed. States of Micronesia	111	6.7	0.7	23	607 islands varying from mountainous to low coral atolls; spread over 2500 km east-west and 1000 km north-south.
Palau	20	3.9	0.5	77	200+ islands; most very small (9 permanently inhabited). 95% of islands and 90% of pop. is within main reef containing Babeldaob, Koror, and Peleliu islands.
Papua New Guinea	6,599	5.5	453.0	13	600+ islands; 80% of pop. in eastern half of island of New Guinea.
Samoa	179	7.6	2.8	23	Two volcanic islands with rugged high mountains and narrow coastal plains: Savai'i (58% of land and 24% of pop.) and Upolu (38% land and 76% of pop.), plus 8 small islands.
Solomon Islands	510	6.3	29.0	18	Nearly 1,000 islands; mixed group of mountainous and coral atolls (350 inhabited). Six main islands account for 80% of land area and bulk of pop.
Timor-Leste	1,098	—	15.0	27	Located on southernmost edge of Indonesian archipelago. Country includes eastern half of Timor island, Oecussi enclave in northwest Indonesian West Timor, and islands of Atauro and Jaco.
Tonga	103	5.8	0.8	25	Archipelago; 176 islands in 4 groups (Tongatapu, Ha'apai, Vava'u, Niua) (36 inhabited).
Tuvalu	12	—	0.026	—	Six atolls with large lagoons enclosed by a reef plus 3 raised coral islands without large lagoons. Funafuti has 22% of land and over 48% of pop.
Vanuatu	233*	5.1	12.0	25	Over 80 mountainous islands of volcanic origin with narrow coastal plains; 90% of pop. on 4 islands.
Total	**9,868**	—	—	—	

Source: World Bank 2008d; SPREP 2004.

Figure A1.6.1 World Bank East Asia and the Pacific Region Member Island Countries

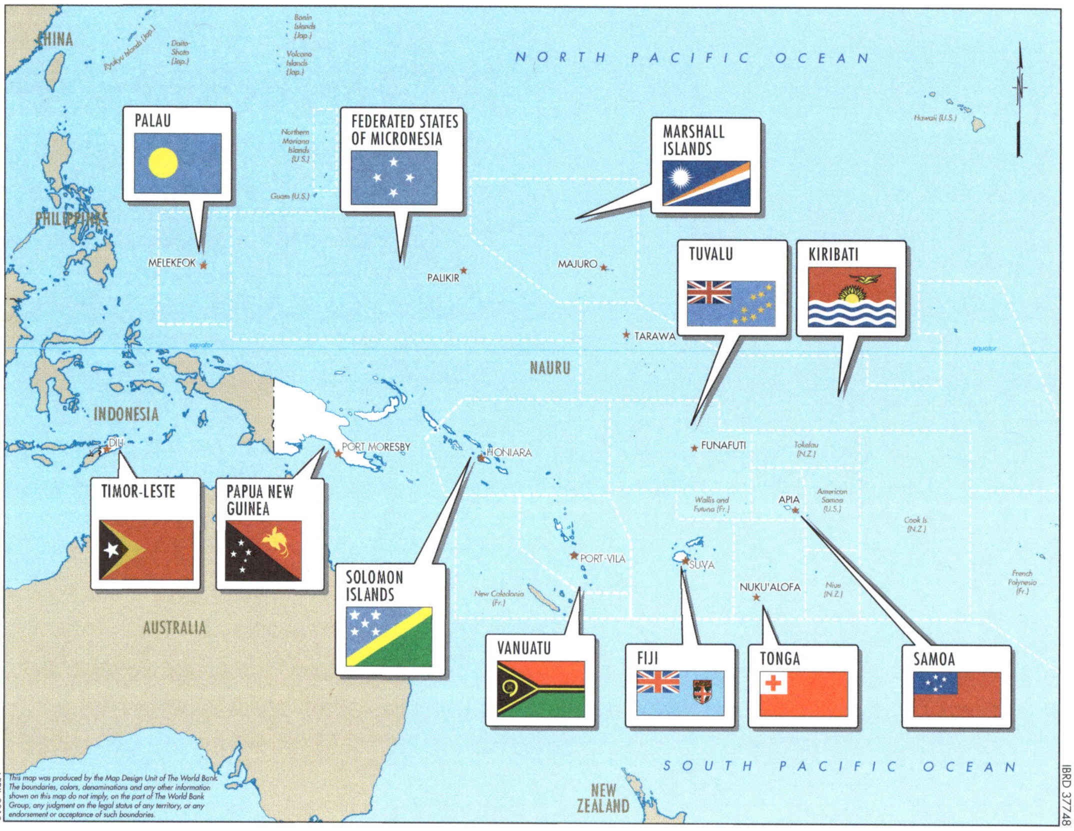

Source: World Bank Design Unit.

local markets typically are low in both supply and demand.

Energy Sector: Extreme Vulnerability to Oil Supply and Price Shocks

PICs are among the highest petroleum-dependent nations. Imported petroleum fuels account for an estimated 99 percent of commercial energy use in the PICs, except for PNG and Fiji. In PNG and Fiji, petroleum fuels account for 60 percent and 20 percent of commercial energy use, respectively (ADB 2009). The 99 percent petroleum dependency rate in the PICs (excluding PNG and Fiji) is the highest in the world. Available sectoral consumption data for the PICs—transport (land, maritime), utility generation, private sector self-generation—are neither reliable nor up to date.[47] Indicatively, in most cases, diesel fuel consumption in PIC electricity grid systems accounts for 30 percent–40 percent of total petroleum imports (approximately 50 percent in the case of Kiribati and 25 percent in Solomon Islands). Also significant, in the vast majority of PICs, the electricity sectors are characterized by nearly 100 percent dependency on imported diesel for power generation.

Extreme vulnerability to oil supply and price shocks. Petroleum dependency makes the Pacific Islands highly vulnerable to oil price shocks—which feed into the affordability of food imports, electricity, and transport. This issue was highlighted in 2008, when world crude oil prices peaked at over US$140 per barrel. However, it should be noted that many Pacific Island countries effectively were

47. In some countries, diesel fuel used for private sector generation can be significant, such as that used by enclave mine operations, resorts, and other enclave developments with self-generation. Such data are not readily identifiable in available statistics.

paying over US$100 per barrel well before then due to the high costs associated with transporting and distributing fuel across the Pacific and among remote islands within each PIC. Many PICs remain exposed due to reliance on spot market purchases of petroleum products with no mitigation measures such as petroleum supply contracts. Those would limit the risks that PICs face from rising oil prices and enable them to capture some of the benefits from falling oil prices.

PICs pay among the highest retail energy product prices in the world. Unleaded petrol and automotive diesel oil (ADO) typically are US$1.30–$1.50 per liter (including taxes) in urban areas but substantially higher elsewhere. The scattered island settlements on the outer islands or in remote small communities on the main islands commonly rely on kerosene for lighting. They pay extremely high retail prices at the end of the supply chain—approximately US$3.00 or more per liter although they generally are not taxed.

Electricity Sector Context

Approximately 8 million people in EAP are in the dark (0.9 million excluding PNG and Timor-Leste). Overall electricity grid access in the PICs is estimated at 59 percent (excluding PNG, in which access is considerably lower). Table A1.6.2 shows the wide variation in coverage levels achieved across the Region. Coverage ranges from near universal access in the Polynesian island nations to under 20 percent for some (Solomon Islands, Timor-Leste) to a low of less than 10 percent in PNG.

Without a fresh start, prospects dim for near-term change. Region-wide, approximately 80 percent of the EAP population lives well outside the reach of existing main grid systems. Approximately 41 percent of the people (excluding in PNG and Timor-Leste) tend to be spread across remote communities of the main islands and outer islands. The human costs of their extreme isolation—from the world at large and within their own country or island—tend to be huge. The ability to participate in the formal Regional economy, communicate in real time, and access basic information and other services to improve their lives is severely limited, if not totally denied.

Indeed, revealed preferences and health statistics point to the two most highly valued benefits derived from even basic electricity access. These benefits enable them to take control of their lives in a way that they are not free to do otherwise. These benefits are (1) communication—the ability to stay "connected" and in touch with people and events outside of their immediate proximate spatial location. Examples of such empowerment are cell phones for voice communications with family and friends, for business and official transactions, and for emergencies[48]; and radio broadcasts for information and entertainment. In addition, (2) at the mere flick of a switch, the ability to turn on a modern—safe and clean—light to read and study for school, extend evening hours, enhance night-time security, and increase income-generation opportunities through both more schooling and later work hours.[49]

Electricity grid systems' performance commonly is plagued by high-cost generation fuel mix and operational efficiency limitations. Electricity grid

48. There has been a rapid increase in mobile telephone use throughout the Region in the last four years, since the introduction of reforms in the countries, which opened their telecoms markets to new private operators. Notably, these countries are Samoa (2006), Papua New Guinea (2007), Tonga (2008), Vanuatu (2008), and Solomon Islands (2010). The resulting competition has caused prices for mobile telephones to fall dramatically. In areas without access to grid electricity, residents have to endure considerable inconvenience in time and costs each time that they get their phone batteries recharged. PIC residents typically pay from US$0.50 cents per recharge in urban proximate zones to $1 per battery charge or even more in remote areas.

49. In particular, the "off-grid population" must make do with grossly inferior alternatives to modern lighting powered by electricity: kerosene lamps, batteries, candles, and flashlights. These options both are costly as a share of typical HH budgets and provide extremely poor illumination. Kerosene lamps in particular pose a serious safety and health hazard, especially to woman and children.

systems in the PICs are operated most commonly by government-owned, vertically integrated utilities.[50] These systems typically serve the larger concentrations of the population in the urban areas of the major towns and the immediate environs as well as the settlements along the higher density coastal belt areas found in some islands. Due to the archipelagic nature of the island nations, a large number of such independent grid systems are operating. They range in size from tiny ("microgrid") systems, with installed capacity on the order of magnitude of 0.1 MW and typically serving approximately 500 customers or fewer to 3 relatively "large" systems. These have generating capacity of 100 MW–150 MW that serves in excess of 125,000 customers each.[51] Most grid systems are in the middle range. Each of 40 such grid systems serves peak demands in the range 1 MW–15 MW. Combined, these mid-size grid systems account for an installed generating capacity of 400MW–500 MW, almost all diesel based. They serve a combined maximum demand of 250 MW–300 MW. Together, the 3 large systems and the nearly 40 mid-size range grid systems among others account for almost the entire 59 percent coverage statistic reported in table A1.6.2.

Financial sustainability a constant challenge for most. Due to a generation fuel mix heavily skewed by diesel fuel as well as operational efficiency standards generally well below efficient practice, most Pacific Island grid system utilities are characterized by a high cost structure. It is not uncommon to find network losses in the 30 percent range. Retail tariffs in the PICs typically range between US$0.35/kWh–$0.75/kWh. Fiji Electricity Authority's (FEA) average tariff, approximately $0.17/kWh, is the lowest due to a nearly 60 percent share of generation from hydro and operational efficiency standards that define Regional best practice. With the exception of Unelco, the private sector utility in Vanuatu, and the public utility, FEA, the utility tariff formulae applied generally do not recover capital costs. Cost recovery would necessitate raising tariffs even higher, when most domestic and small business customer consumption already is severely limited due to the lack of affordability of electricity bills in relation to HH budgets.

For most PIC utilities, the dream of achieving financial sustainability is oriented at best toward an initial big step of achieving recurrent cost recovery via retail tariffs. The ability to do so means that utilities need to begin with a proper least-cost plan and large-scale investment in diesel fuel substitution. This substitution should go from grid-connected, cost-effective renewable energy options—hydro, geothermal, and solar PV—and investment to lower network losses and the specific fuel consumption of existing diesel stations. However, donors generally have not supported efforts to maintain infrastructure as an economic and productive asset. Furthermore, the practice has been that, rather than being systematically planned and implemented at the optimal times, new investment is rare. Customarily, planning is not the driver of new generation and network investment timing and of the type and specifications. Rather, investment tends to be wholly grant financed and off balance sheet. Moreover, its timing usually depends on the whims of donors, most crucially, when an interested donor may step forward to invest to keep the lights on. These actions typically occur in response to a looming crisis, with the donor specifying the choice and type of equipment.

50. An exception is the private utility, UNELCO, in Vanuatu. In Tonga, the private utility, Shoreline, was sold back to the state in 2008. The 100% state-owned successor, Tonga Power Ltd (TPL) has a regulatory contract that specifies a 12.9% return on regulated assets and, in principle, should be able to cover capital costs. In Fiji, a 9 MW biomass-based IPP sells to FEA and plans call for more IPP development. In PNG, entities other than PPL increasingly are generating power. Such entities include provincial-government-operated small grid systems and mining enclave operations and church organizations that supply and distribute power to adjacent communities.

51. Fiji Electricity Authority (FEA)-operated grid systems on the main island of Viti Levu have installed generation capacity of 185 MW. PNG Power, Ltd (PPL) Port Moresby grid system has an installed capacity of approximately 140MW, and the Ramu grid system has 200MW of installed capacity.

Renewable energy: Weak record due to misdirected efforts. Starting in the early 1990s, donors directed substantial interest and support to increasing the use of renewable energy technology in the energy supply mix, almost as an end in itself. This interest shows prominently in the form of numerous studies and project activities in the Region.[52] In the absence of national or utility strategic planning to indicate priorities, donor financing has been largely misdirected, driven by funds earmarked for priorities developed at a global scale rather than for PIC-specific priorities. A disproportionate amount of funding has been directed to off-grid renewable energy technologies for access and otherwise driven by climate change, which is a developed country agenda.

National energy infrastructure investments and related support for the main grid systems largely have been neglected. Furthermore, even in off-grid areas to which investment has been delivered, the results and performance do not measure up. For example, in most countries, solar home systems (40Wp–100Wp range) have yet to achieve measurably significant penetration areas.[53] Furthermore, instances of renewable-energy-based micro/small grids that have been running with a track record of good performance and maintainability as well as cost competitiveness are hard to find. Donors generally have focused more on demonstration and testing, and less on achieving efficient costs, reliable performance, and scalability.

52. A notable exception is Fiji, which has aggressively developed hydropower resources; has a power purchase agreement with a biomass-based IPP; and has an active program to further promote expand cost-competitive, renewable-based generation supply. Vanuatu's promotion of coconut oil so far has been uneconomical.

53. One notable exception is Kiribati's outer islands electrification program. It has been operated for several years by a government-owned entity, Kiribati Solar Electricity Corporation (KESC). With grant financing of its capital costs by the European Union, KESC operates a fee-for-service-based SHS program on the outer islands for homes, community halls, and key institutions.

Energy Sector: Main Policy Challenges

Managing energy security risks: Continuing key role of hydrocarbons. The petroleum supply chains to the Pacific Islands are long, narrow, and vulnerable to disruption. Increasing the share of renewable energy will reduce the extent of oil dependence and the risk of supply disruptions as well as reduce the economic vulnerability of these countries to oil price shocks. However, for the foreseeable future, imported petroleum fuels will continue to play a major role in energy supply and demand, and not only in the transport sector. To reduce the growth of demand for imported fuel and to reduce costs and cost volatility, improvements in the petroleum supply chain and in price risk management, as well as improved customer efficiency, must be priorities. These improvements will require significant investment in new assets in the petroleum supply chain and in the electricity-generation sector. Just as urgently, improvements, including investment, are needed in safety and emergency response systems and readiness. However, most Pacific Island governments are suffering severe budget pressures from shrinking income sources, and rising costs, one of which is for oil. Tight budgets restrict possibilities for investment in initiatives for oil substitution or efficiency improvements. Meanwhile, as noted above, development partners have tended to direct their support to off-grid activities.

Addressing climate change: Putting people first. The global nature of greenhouse gas (GHG) emissions puts pressures on all nations to reduce their carbon footprints and at times introduces conditions for access to external funding sources that have more to do with global trends than national priorities. Nevertheless, even if access to fossil energy increased significantly across the Pacific Islands, their emissions would add negligibly to global GHGs. The challenge is to formulate and implement a balanced set of energy policies and initiatives. These can be identified through a credible least-cost planning process that highlights the expanding role of low-carbon energy forms while recognizing that petroleum will continue to play a key role in meeting PIC energy needs.

The Regional energy agenda is framed appropriately as one to improve energy security rather than to promote renewable energy. The latter agenda is driven by global climate change concerns. Development of the renewable energy supply is an important part of the diversification strategy to improve energy security. Specifically, increasing the share of renewable energy—on- and off-grid—should not be seen as an end in itself nor even as the sole effective route to reduce GHG emissions. Rather, increasing the renewable energy share is one among several means to be deployed and assessed within a balanced, bigger-picture approach to energy issues. The ends sought are to ensure secure, affordable, widely accessible, high-quality energy services to the people of the Pacific.

Strategic Directions: New Horizons and Opportunities

This section draws attention to relevant strategic and policy directions to address the interlocked set of challenges to secure and affordable energy access for the citizens of the Pacific Island Countries. The key messages that follow are intended to inform the approach, scope, and depth of country-specific strategies and policies; and the staging of actions and specific investments tailored to the PICs' respective national priorities and sectoral contexts.

20-20 vision of the problem definition: "An Energy-Secure Pacific." Pacific Island nations face a tangled web of challenges and constraints to the formulation of a coherent set of national and regional energy policies and actions coordinated across key sectors and stakeholders. The 2010 endorsement of the "Framework for Action on Energy Security in the Pacific" was a significant milestone and a big first step in the right direction.[54] This framework recognized that the common thread weaving throughout the energy sector challenge and those related to addressing energy access is the PICs' extreme vulnerability to high oil dependency, high import costs, and supply and price shocks. Accordingly, the framework formulated a clear and concise vision of an "energy-secure Pacific":

> "Energy security exists when all people at all times have access to sufficient sustainable sources of clean and affordable energy and services to enhance their social and economic well being."
>
> —Framework for Action on Energy Security in the Pacific, 2010

Bringing it all together: "Whole-of-sector" framework and principle of "many partners, one team, one plan." Besides clearly articulating the problem definition and development outcome sought for the energy sector, the forum leadership rightly signaled the imperative to break from the past. This shift can be characterized as a move away from a fragmented "project-by-project and donor-by donor approach" toward a programmatic approach. The latter will be embedded in a sector-wide framework and process that is country owned and country driven with joint accountability for results that are aligned with national priorities. A notable and example-setting application of the whole-of-sector approach and organizing principles is the recently completed Tonga Roadmap (box A1.6.1).

Grid Electricity Systems: Time for Strategic Investment in Renewable Energy

Substituting diesel fuel in grid-based electricity generation arguably would be the most effective way to reduce the sector's petroleum dependency. Substitution could be achieved through improving energy efficiency and cost-competitive and proven

54. "Toward an Energy-Secure Pacific: A Framework for Action on Energy Security in the Pacific (2010–2020)" (SPC 2010). This framework is the outcome of an extensive and widely cast consultative process coordinated by the Secretariat of the Pacific Community (SPC) (2010). Extensive consultations took place with the PICs; regional power utilities; private sector; nonstate actors; major regional development units at other organizations; bilateral and multilateral development organizations; and development partners from Australia, China, Italy, Japan, and New Zealand, among others. On June 30, 2010, a special forum of energy ministers was convened in Brisbane as part of the lead-up to the Forum Leadership summit held August 5, 2010 in Port Vial, Vanuatu.

Box A1.6.1 Tonga Energy Roadmap: Many Partners, One Team, One Plan

> "Energy is a fundamental building block for the Kingdom in its social and economic development and in enhancing the livelihood and well-being of all Tongans. It affects all businesses and every household. Accessible, affordable and sustainable electricity that is environmentally responsible and commercially viable is a high priority. My government recognized the importance of having dependable, accessible and reasonably priced power as a key catalyst for sustainable economic growth. Achievement of these goals is crucial to achieve the government's primary target of 'poverty alleviation' including 100 percent accessibility to electricity."
>
> —Hon. Dr. Fred Vaka'uta Sevele, Prime Minister of Tonga, April, 2010

The "Tonga Energy Roadmap" (TERM) effort was led by the government of Tonga (GOT) in coordination with the World Bank. The effort convened development partners and embarked on a sector-wide review to improve the performance of the energy sector and to mitigate the risks. In so doing, TERM recalled and redefined a sector-wide framework and process that colloquially has become known as a "whole-of-sector" approach to energy planning and investment.

Development of TERM involved unprecedented information sharing by government ministries and Tonga Power Ltd (TPL). The process set a new standard in the Pacific for government leadership and coordinated development partner support. Over the past year, the process has benefited from the active participation of more than 15 development partners, including multi- and bilateral agencies and regional organizations responsible for energy in the Pacific. TERM is important not only for its expected impact on the Tongan energy sector. Other PIC energy sector discussions already are citing the roadmap as exemplary and citing the government's leadership and development partners' coordination of sector-wide planning applicable to infrastructure more broadly.

The roadmap covers 1 decade starting in 2010. It addresses improvements in the petroleum supply chain and considers price-hedging instruments. These comprise reducing cost and volatility in the price of petroleum imports and delinking electricity production from petroleum to the extent feasible. TERM also focuses on (1) increased efficiency in electricity supply and use, (2) development of grid-connected domestic renewable energy resources, (3) improved access to quality electricity services in remote areas, (4) reduced environmental impacts both locally and globally, (5) enhanced energy security, and (6) overall sector financial viability. TERM's scope also includes policy, legal, regulatory, and institutional aspects of the sector as well as investment. It is envisioned that TERM will be updated periodically to take into account technologies, costs, demand for electricity, and sources of financing as they evolve.

GOT formally adopted TERM in August 2010. The World Bank is providing GOT ongoing support to implement TERM via the Tonga Energy Roadmap Project, which is under preparation, The project will reinforce the focus on, and provide the support to achieve, the medium-term energy sector outcomes highlighted for this proposed operation. Other development partners also are working with GOT to prepare support aligned with specific aspects of TERM implementation.

renewable energy-generating options: hydro and geothermal power generation. Diesel substitution would be followed by efficiency improvements on the utility and customer sides.[55] Moreover, grid-connected large-scale solar PV generation increasingly is being deployed effectively in island contexts all over the world. Examples include remote operations on several Caribbean basin island utilities, Reunion Island in the Indian Ocean, and, more recently, in EAP.[56] Present-day good-practice technical designs and component

55. Estimated by the scale of efficiency improvements relative to the timeframe per unit of investment.

56. New Caledonia: 2.1 MW grid-connected central power station, operational since April 2010; Tuvalu: 40 kW solar PV power station.

specifications for key equipment components, such as inverters, can effectively address performance and maintenance issues in the earlier generation installations, even in remote settings. Nevertheless, until recently, the deployment of such designs in the PICs' grid systems usually was absent.

For many PICs, hydro, geothermal, and solar-PV-grid-connected generation—grid systems—represent the largest scale, fastest, and most cost-effective interventions for increasing access to electricity. More specifically, in the Pacific Island context, electricity generated from utility-scale, grid-connected solar power stations already is competitive with today's oil prices. On both a total-cost basis (26USc/kWh–28c/kWh versus 30c/kWh–35c/kWh), and a variable-cost-only basis, solar PV generation is far more cost-effective than diesel (1c/kWh–1.5c/kWh versus 25c/kWh–30c/kWh). Looking ahead, unit costs of module production are expected to decline further along the "technology learning cost curve" in response to the rapid global increases in the annual capacity of installations of solar PV module capacity.[57] As this decline continues, the cost advantage in favor of solar PV generation will continue to increase. However, it needs to be kept in mind that the amount of solar PV power that can be added and its impact on total cost is limited by its intermittent nature without the means for storage.

After hydro and geothermal, grid-connected solar PV generation is the most versatile and dependable among the intermittent supply options available and deployable across the Pacific Islands. Over time, PIC utilities can significantly reduce electricity generation costs and improve financial performance. These results can be achieved through a well-staged and systematic implementation plan that duly considers efficiency improvements on both the supply and customer sides. In this way, the utilities can lower subsidy burdens on their governments, while over time giving some relief on electricity tariff levels.

57. Grid-connected solar PV has grown by an average of 60% every year of the past decade, increasing 100-fold since 2000 (Ren21 2010).

Grid Systems Target Two-Part Renewable Energy Goal

Diesel conversion/substitution program*: *20 percent–25 percent substitution of diesel generation. As discussed above, the EAP Region hosts 42 grid systems: approximately 40 mid-size grid systems and 1 large system each on Fiji's main island (FEA) and Port Moresby. Combined, these mid-size and large grid systems account for a *peak load* of approximately 450MW and installed generation capacity of 700 MW[58] (of which at least 600MW is diesel based). As a reference point, a first-stage programmatic goal of substituting at least 20 percent of today's diesel generation share over a 10-year horizon through cost-competitive renewable energy[59] is realistic.[60]

A separate goal of 50MW of grid-connected capacity from renewable and other sources including solar PV, wind, and biomass is an indicative target to be phased in over the medium to long term.[61] Early-stage systems would be designed simply to supply energy as available. Over time, these systems increasingly would use battery storage systems to an extent for firm power availability

58. Additionally, PNG has captive power generation, mainly for the mining industry, of approximately 300MW.

59. Where available, hydro or geothermal energy; otherwise, primarily solar PV and, in some instances, biomass or wind energy.

60. Outside of PNG, electricity demand growth is projected to be low in the short to medium term. As a consequence, for most PIC grid systems, the renewable energy generation investments will be driven by a "conversion/diesel substitution" program. The program's objectives will be to (a) lower the costs of electricity generation and tariff relief for electricity users, and (b) diversify fuel to enhance energy security via reduced supply risk.

61. Fiji already has established a diversified generation mix and has additional plans to diversify. PNG's considerable hydro and other resources tend to be the dominant prospects. The Solomon Islands have good hydro prospects.

during periods of peak demand ("peak shaving"). Along the way, PIC utilities will gain operating experience in integrating intermittent systems—with and without some storage (see below)—and in building institutional capacity. These developments will set the foundations for those interested in deploying even "deeper penetration" of grid-connected solar PV capacity that uses sophisticated control systems hardware and software to manage the network operating system.

In this future context, a relevant development is that battery storage technologies are making it increasingly feasible and attractive to use solar PV power as a dispatchable resource for peak shaving. A case in point is sodium sulfur (NaS) battery technology, which has been proven over 2 decades in utility-scale applications. This technology was applied in the Tokyo Electric Power Company grid system (200MW of NAS batteries); in northern Japan (51MW hydro station coupled with 34MW NAS battery system); and more recently in Texas (4MW battery). In addition, use of the NAS batteries in a remote island context is undergoing ongoing testing and measuring over 2010–12 in Reunion Island in the Indian Ocean. However, the investment and operating costs likely will remain a major constraint in the near future versus deployment in the PICs (box A1.6.2).

What is most important are to (1) get supporting financing and TA mechanisms in place to enable a Region-wide push, and (2) organize and implement a systematic programmatic framework. Once the program is underway, timeframes can be better calibrated as they are informed by evolving implementation and operation experience. The first-phase buildup of grid-connected renewable energy generation investments also will help set the stage for longer term, least-cost expansion plans.

Operational Efficiency Improvements: Capacity Building and Upstream Technical Assistance for the Grid System

For many PIC utilities, a top priority is first to get their own houses operationally in order. Among the highest priority interventions are implementing network loss reduction programs (many PICs have system losses in the high 20 percent–low 30 percent range[62]), reducing specific fuel consumption (SFC) of diesel power stations, improving billing and revenue management practices and systems, and sound advance planning. These interventions will need ongoing investment. Also crucial will be dependable and relevant TA as well as strengthened institutional capacity to diagnose and scope, design, and manage ongoing specific operations improvement programs. In this context, the Fiji Electricity Authority (FEA) is considered the Region's best-practice utility (box A1.6.3).

Using Simple Power Solutions to Change Lives of Poor and Remote Dense Populations: Thinking and Acting Differently

Approximately 8 million people in the PICs live in off-grid areas (6.6m in PNG, 0.9m in Timor Leste, and approximately 0.5m in the other PICs). Most of these people are unlikely to receive grid service in the foreseeable future.[63] Conventional solar home systems (SHS) (that power 2–3 lights and a black and white TV) are not affordable for most poor HH. Typically, these systems cost at least US$400–1,000+ (2–5 times the per capita income of most PICs) coupled with US$10–$15 monthly service fees for the huge costs of supporting extended and specialized service and maintenance chains. Furthermore, with the exception of Kiribati (mentioned earlier), SHS have not achieved significant market penetration.

A recent generation of pico ("very small") solar charger consumer products is available for the most essential life-giving elements that have been denied to remote and poor populations. These

62. Systematically collected and up-to-date reporting of data is vital to inform the diagnostics for each grid system because key performance indicators—technical, commercial, and customer service—are not yet readily available.

63. In the long run, it is projected that approximately 2.2 million of the 8 million are likely to be served by grid systems in PNG and Timor Leste.

Box A1.6.2 Advanced Battery Storage Technologies Enhance Integration Potential of Intermittent Renewable Energy Generation into Grid

Solar, and more so, wind energy generation are inherently intermittent. This reality limits the typical extent to which such generation can be absorbed by island-scale grid systems to under 15 percent–20 percent. This limitation is determined by several operational considerations. These derive from the technical imperative that requires that supply and demand in a grid system be matched instantaneously and ongoing. Specifically, in the absence of storage, intermittent generation reduces power quality (harmonics and voltage fluctuations) while it increases the need for balancing in real time. This balance is achieved by increasing the reserve firm capacity on line to maintain the grid system frequency to rated 50Hz. However, when coupled with proper storage, otherwise intermittent renewable energy from grid-connected solar PV power stations and wind can be stabilized as well as counted on as a firm dispatchable resource to balance system demand and supply at each instant.

NaS Battery Storage System

Over the years, many battery technologies have been proposed and developed for electrical energy storage applications. Nonetheless, only a handful have been used in field systems and proven on a utility scale. A promising case in point is the sodium-sulfur (NaS) battery storage system. NaS batteries have a relatively small footprint (high energy density); high discharge efficiency (89 percent–92 percent) if used regularly; integral thermal management; and low maintenance requirements. NaS batteries also offer cycling flexibility and long cycle life, and are fabricated from inexpensive materials.

Box figure A1.6.2.1 Use of NaS batteries for peak shaving

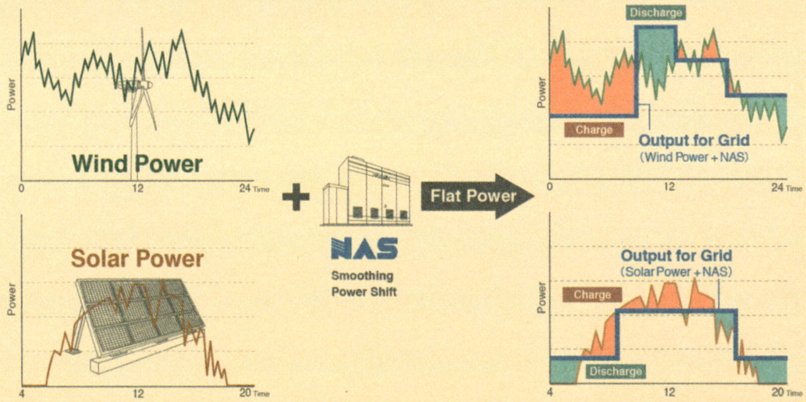

Source: http://www.jase-w.eccj.or.jp/technologies/pdf/electricity/E-18.pdf

Currently, NaS battery costs are approximately US$2,500 per kilowatt—approximately 10 percent more than the cost per kW of a new coal-fired plant. Mass production is expected to drive down NaS battery prices.

Reunion Island Grid System Installation of a 1MW NaS Battery

Since 2009, an NaS battery system has been piloted in France's Reunion Island in the Indian Ocean to provide firm peak shaving power on the island grid system. The installation is integral to a grid-connected solar PV power station, and is rated for 7 hours of firm power per day and for 2,500 cycles over its lifetime. The battery supplier guarantees a 15-year life and performance, and provides turn-key scheduled maintenance per design and performance specification. The NaS battery system requires scheduled maintenance every 3 years, and the entire battery module to be replaced every 15 years, with recycling included.

Sources: http://www.ngk.co.jp/english/products/power/nas/index.html; Electrochemical Society Interface 2010; Drouineau and others 2010.

Box A1.6.3 Fiji Electricity Authority: Good-Practice Neighbor Offering a Helping Hand

The Fiji Electricity Authority's (FEA) record of accomplishments and performance is exemplary. FEA's record not only stands out in the context of Pacific Islands national utilities but also places it among the best of island nation utilities globally.

Beginning in 1966, FEA had a generating capacity of 3,394 kilowatts and only 2000 customers. The utility has grown steadily and now provides electricity supply via several grid systems to approximately 150,000 customers in the 3 main islands of Viti Levu, Vanua Levu, and Ovalau (81 percent coverage overall). Today, the installed generating capacity stands at 205MW, comprising a combination of hydro, a wind farm, and several diesel power stations (100MW). An independent power producer (IPP) supplies electricity to FEA via biomass generation.

Noteworthy examples that serve to highlight FEA's impressive and sustained efficiency gains and serve as reference benchmark beacons for PIC utilities follow.

System Losses at Best-Practice Levels

The moving average system loss—network losses and auxiliary station usage—for the last 5 years ranges from 8.70 percent to 10.23 percent. These numbers compare favorably with the international benchmark of approximately 10.00 percent.

Box figure A1.6.3.1 FEA staff explain features of new electricity connection to customers

Photograph: Fiji Electricity Authority.

Diesel Station Fuel Performance

FEA operates 9 thermal power stations with an installed capacity of 110 MW. The average specific fuel consumption (SFC) across all these diesel stations is 0.2419L/kWh with the station usage at 1.84 percent of total generation. This good-practice performance is underpinned by effective work practices and discipline. These include regularly scheduled maintenance on all fuel injection systems, tuning fuel pump timing, ensuring loadings of the generator sets for optimum fuel efficiency, and attending to fuel supply quality management. FEA recently signed a Fuel Purchase Agreement with an international supplier. This agreement includes a value-added initiative undertaken by the supplier: implementation of improvements in fuel consumption of the diesel generation sets. This initiative will lead to a minimum additional 1 percent fuel efficiency and savings.

Throughout the 1990s through 2008, FEA's average tariff level stayed nearly flat in nominal terms at approximately 21FJc/kWh–22 FJc/kWh. In 2009 the tariff rose to 25.5FJc/kWh (still well below inflation). In June 2010, the Commerce Commission approved a price increase that raised the average to 32.3 FJc/KWh. In comparison, tariffs in Australia and New Zealand are 42.4FJc/kWh and 47.5FJc/kWh, respectively. Apart from Fiji and Palau, tariffs for all other PICs range from 60FJc/kWh (PNG) to 147FJc/kWh (Solomon Islands) (box figure A1.6.3.2).

This tariff story is a proxy bottom-line testament to FEA's management- and results-driven culture. This culture is driven by technical excellence, professionalism, and accountability instilled throughout the organization; orchestrating a sustained and impressive record of cost containment and productivity gains, while continuously expanding and improving service and maintaining financial soundness.

(continued)

Box A1.6.3 Fiji Electricity Authority: Good-Practice Neighbor Offering a Helping Hand *(continued)*

Box figure A1.6.3.2 Electricity tariffs among lowest in Pacific, including Australia and New Zealand

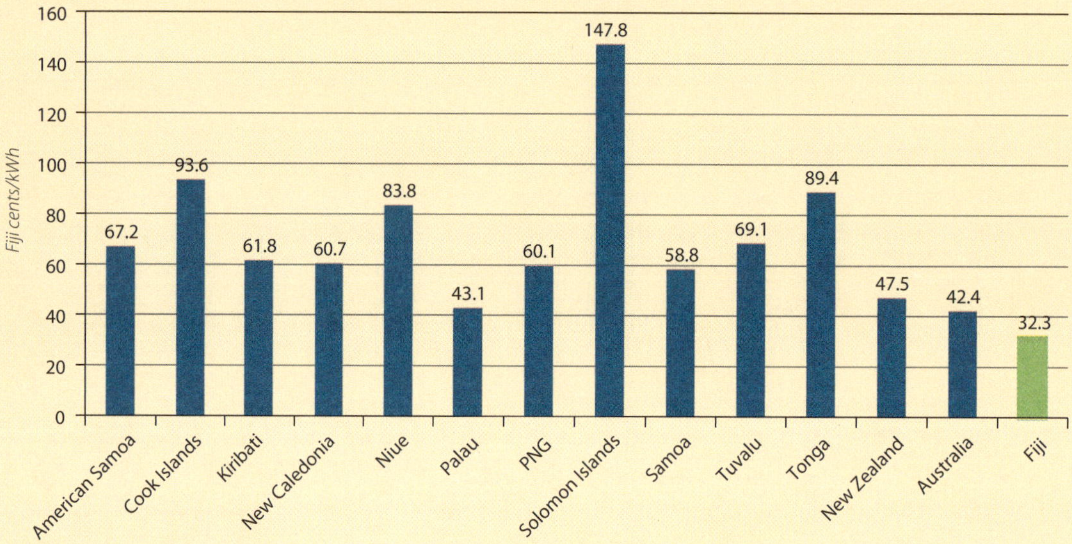

Source: FEA Statistics 2010.

Training Institute in Lautoka

FEA also operates a well-run facility for training front-line technical personnel in Lautoka.

In sum, FEA is a proven Regional resource that PIC utilities can readily tap for knowledge, TA, and upstream diagnostics under arrangements such as twinning with, and secondment to, FEA. FEA experts also can visit utilities that require assistance at the front end of their process to advise and scope.

Source: FEA statistics and discussions with FEA management team 2010.

devices are powered by a small solar PV panel. They contain a built-in battery that can store sufficient electrical energy to charge cell phones and small radio batteries, and power a light-emitting diode (LED) light. Pico chargers' small size and portability have been made possible by the development of LED bulbs, which are 30 percent–50 percent more efficient than even high-efficiency compact fluorescent lamps (CFLs). Conveniently as well, LEDs use DC power, which is the output of a solar panel. In contrast, CFLs use AC power, which requires an additional inverter component, thereby adding to CFLs' cost and complexity.

Rapid Scale-up Potential with Genuine Global Reach. The better performers among these solar chargers are user friendly and have robust circuitry for long-life performance to withstand high heat and humidity and rough use. These solar chargers do not need special-purpose and high-cost supply and after-sales maintenance chains, as do SHS.[64] Instead, these products are characterized by cash-and-carry, plug-and-play features. Hence, they are readily piggy-backed on established retail trade supply chains and other existing networks, such as those of rural NGOs, church organizations, and cell phone service operators, that extend out to Regional and localized markets commonly found in the PICs. These small solar-powered devices

64. These "plug-and-play" devices can be replaced at relatively low cost. In comparison, for people in remote rural areas, SHS require constant and ongoing maintenance by the users and servicing chains.

offer the real prospect for rapid market penetration within a decade (box 3.5).

Right push needed at this early "starter-market" phase. In the formative stages, government's roles include enabling market development by helping promote product awareness, providing reliable information on product(s) quality and price, providing incentives to improve product quality, and enabling wholesale trade finance. Also vital at this early stage is a modest capital subsidy.

Case for price subsidies in starter-market development phase. Specifically, under today's conditions, to achieve retail prices for pico solar devices that range between US$10–$25 depending on the features, price subsidies are required.[65] Such subsidies can range from 35 percent to 60 percent (for high-end phone chargers), equivalent to subsidies of $20–$35, respectively. Eventually, it is expected that today's better performing devices will rapidly achieve high market scale at prices affordable to the majority of beneficiaries.

Delivering better value for money: Partnerships and funding mechanisms that align with national and Regional priorities. Above all, PICs need financial assistance to formulate realistic plans to expand grid-renewable energy and improve operations. Funding should align with national and Regional priorities and avoid being tied to specific technologies. Past funding decisions too often were driven by donor priorities or global agendas. These priorities and PICs' needs frequently are not the same.

Successful implementation of the strategy and wide-ranging actions underpinning the vision of an energy-secure Pacific hinges on:

- The degree of success achieved by PIC governments in fostering and leading effective working partnerships across key Regional institutions
- Leveraging resources from development partners, and possibly from "special global funds," for jointly agreed Regional and national investment and TA plans aligned with Regional and national priorities.

A noteworthy initial step away from the donor-by-donor and fragmented approach was established with the formation in 2008 of the Pacific Region Infrastructure Facility. PRIF is under the joint leadership of Australia, New Zealand, the Asian Development Bank, and the World Bank. The facility emphasizes better linking of investments to longer term reforms in the way that Pacific Island countries plan, manage, and finance their infrastructure needs. In addition, PRIF provides an effective mechanism to marshal coordinated donor support for investments in economic infrastructure and to address the problem of funding for infrastructure asset maintenance.[66] PRIF has a way to go to achieve the goals of harmonization, alignment, and coordination across a wider set of donors. However, formation of this facility does mark a promising significant step in the right direction.

Enhanced expectations for key Regional institutions: The "many partners, one team, one plan" approach. Finally, progress in implementing the framework for an energy-secure Pacific will require harnessing and exploiting to the fullest the synergies arising from the respective

65. PIC governments and donors have been subsidizing new connections for grid and off-grid beneficiaries all along. These capital investment subsidies arguably exceed $1,000 per grid connection and at least $500 per off-grid SHS connection plus subsidies for overhead costs of the SHS program.

66. The following are examples of ongoing PRIF efforts, especially those involving Australia and World Bank, that capture the reasons underlying several of the strategic directions outlined in the preceding section. They are (a) Tonga: Energy Roadmap (support for some elements of the preparation); (b) Kiribati: Grid-connected solar PV power generation—central station and distributed—for diesel fuel substitution of the S. Tarawa grid system (investment, TA, and project preparation); and (c) Vanuatu: (1) Efate Geothermal Power and Island Ring Grid Development Program and Framework (pre-feasibility study) and (2) Hydrocarbon Supply Chain Management Efficiency Study.

mandates and comparative strengths of the key relevant regional institutions and organizations.

An example would be a grid-connected renewable energy program mounted along the lines outlined above. There is a clear, urgent need for a collaborative and coordinated program going forward. This program could involve a joint working partnership of the Secretariat of the Pacific Community (SPC), Pacific Power Association (PPA), World Bank, and other interested partners. The purpose of the partnership would be to monitor and assess in real time the on-the-ground experience with grid-connected renewable energy in the PICs (as well as in relevant countries outside the Pacific). This partnership could feed back pertinent lessons learned to countries that are trying to increase grid-connected renewable energy and strengthen their capacities to chart their courses in that direction.

A1.7 Vietnam

The rural electrification effort in Vietnam has been a remarkable achievement. The share of HH with electricity access grew from 2.5 percent in 1975 to 96 percent by 2009 (World Bank 2011b). Through an unparalleled effort, the country provided access to more than 80 million people over 33 years. Vietnamese with access to electricity increased from 1.2 million in 1976 to approximately 82.0 million in 2009.

Along the way, the government addressed a wide array of challenges. For example, it successfully balanced the sometimes-competing interests of local, provincial, and central governments; government programs; and, in later years, support from

Population *(2008) (mil)*	86.2
Rural population *(% total population, 2008)*	72
Population density *(people/sq km)*	278
Land area *(sq km)*	310,070
GNI per capita *(Atlas Method: Current US$, 2008)*	890
Access to modern cooking fuels	34
Urban *(% HH, 2008)*	76
Rural *(% HH, 2008)*	20
Electricity access, national *(% HH, 2009)*	96
Urban *(% HH, 2008)*	99
Rural *(% HH, 2008)*	94
No. of people w/o access to electricity *(2009) (mil)*	4–5
Population served by off-grid sources (minigrids and HH systems) *(%)*	1
Electricity access target and year *(% HH)*	100; 2020
Electric power consumption *(kWh per capita, 2007)*	728
Installed capacity *(MW 2008)*	13,850
Thermal	8,350
Hydro	5,500
Other renewable	—
Electricity net generation *(bil kWh)*	69,965
Distribution losses *(% net generation)*	10.5
CO_2 emissions *(M/T per capita, 2007)*	1.3
Indicative residential electricity tariffs for rural consumers *(2011)*	USc3.1/kWh; EVN customers, first 50 kwh @600 VND/kwh *(Source:* EVN)
Key institutions for electricity sector	Electricity of Vietnam (EVN) Electricity Regulatory Authority of Vietnam (ERAV) Ministry of Industry Distribution Power Corporations (PCs)

Sources: World Bank 2010e; www.eia.doe.gov; IEA 2010; WHO and UNDP 2009.

Vietnam has a diverse topography with steep mountains in the north and flat coastal lowlands in the south. The rural areas are densely populated. The poorest communities are found in the more remote areas of the central highlands and northern mountains, in which 75 percent of the country's ethnic minority populations reside.

Table A1.7.1 Vietnam: Scenario Analysis for Universal Electricity Access by 2030

	Total investment needs (US$ mil)		Incremental needs for universal access (US$ mil)			
Period	**Business-as-Usual scenario**	**Universal Access scenario**	**Total**	**Urban grid**	**Rural grid**	**Off-grid**
2011–20	1,459	1,459	—	—	—	—
2021–30	1,064	1,064	—	—	—	—
2011–30	**2,522**	**2,522**	—	—	—	—
		Annual requirements:	—	—	—	—

Note: No. of HH without electricity in Business-as-Usual scenario by 2030: <1%.

development partners, to create an institutional structure for rural electricity supply that bodes well for long-term sustainable development.

Over time, Vietnam's rural electrification (RE) program and its priorities have evolved. The issues encountered during the RE effort were constantly changing and the challenges daunting. At the outset, the country did not have one unified master plan. The government continually evolved in the ways that it tackled the challenges, which themselves were shifting. The challenge of the early stages—to provide simple connections—evolved into one of securing quantity and quality of supply combined with meeting ever burgeoning urban and rural energy demand.

Progress of Rural Electrification in Vietnam

Vietnam's electrification progressed through several phases, or periods, to its present level:

Recovery period (1976–85). After three decades of war, with underdeveloped and isolated systems, electricity supply was available only for cities and large industries. "Productive uses" were given priority. The HH electrification rate grew from 2.5 percent to 9.3 percent.

Preparation period (1986–93). The Doi Moi reforms were implemented.[67] They resulted in an increase in rural incomes, the development of several large power plants across the country, the start of the construction of the 500 kV transmission line running the length of the country, and the building of medium voltage networks. The foundations laid by these important developments enabled significant progress in rural electrification in the subsequent periods.

"Take-off" period (1994–97). Electrification experienced remarkable growth. The HH electricity access rate more than quadrupled from 14 percent to 61 percent. The local and central government authorities responded to the strong societal demand for electricity. Indeed, by the mid-1990s, electricity access rates had become one of the key indicators in the annual socioeconomic development assessment of every commune, district, and province. Meeting the increase in demand for electricity access had become possible as a result of the completion and coming online of two significant sources. They were (1) the last unit of the Hoa Binh Hydropower station as well as other relatively large power plants; and (2) the 500 kV transmission line, which made abundant power sources available throughout the whole country. Other important developments during this period were the establishment of Vietnam Electricity (EVN) in 1995 and the clear nationwide electrification targets set by the government in 1996. During this period, the government and the World Bank began preparing their first joint energy sector project.

Focus on regulation (1998–2004). The growth in electricity access rates over these 7 years from 61 percent to 87 percent of HH was slower than in the previous period. Instead, this period was host to a series of fundamental changes in the sector that would pave the way for future sustained success. A significant feature was more pronounced government involvement in determining the course of rural electrification through three means. They comprised (1) defining strategies for the planning, implementation, and management of rural electrification; (2) setting the legal framework for the sector; and (3) engaging with Vietnam's international partners to implement its strategic priorities. Two important actions that resulted were Decision 22 and Decree 45, which set out institutional and financial arrangements for the electricity system.

This period also witnessed the passage of the country's first electricity law and the issuance of a policy paper on rural electrification. Another critical government initiative was to set a ceiling tariff for rural customers as a step toward establishing financial controls over the rural electricity supply business. Technical specifications for rural electricity systems also were developed, and were formally adopted in later years.

During this period, institutional arrangements and sources of financing for rural electrification shifted significantly. EVN began a pilot program for operating low voltage (LV) systems and started to acquire medium voltage (MV) systems that had been financed, by other entities. Under the service agent model, created under the First Rural Energy Project, local community members maintained LV systems on behalf of the PCs, carried out simple repairs, and handled collections. This model helped ensure accountability within local communities, minimize nonpayment, reduce system losses, and significantly

67. "Doi Moi" refers to the economic reforms initiated in Vietnam in 1986 with the goal of creating a "socialist-oriented market economy."

lower the costs of system operations and management (O&M) for the PCs.

Focus on quality and regulation (2005–09). Meanwhile, the quality and reliability of electricity supply were emerging as important issues. The government chose to pay attention to these issues and to continued expansion of electricity access. This period could be characterized by the enforcement of legal requirements; a shift in focus from network extension to rehabilitation; and direct government support for extending electricity access, particularly to minorities and those in remote areas. The government's focus was not only on increasing electrification rates but also on ensuring efficiency and addressing institutional shortcomings in the sector.

Consolidation for the last mile (2009). The period from 2009 onward was characterized by solidifying the focus on ensuring the sustainability of the rural electricity supply business. Simultaneously, the government was pushing for greater accountability, working to determine the most appropriate strategies for extending access to those without electricity, and ensuring the affordability of electricity for the poor.

An important milestone of this period was the Prime Minister's Decision 21, issued in February 2009. It stipulated a unified tariff for all consumers, along with an incremental block tariff arrangement, with a new lifeline block. The decision also enabled the takeover of financially weak local distribution utilities (LDUs) by the PCs, now renamed Power Corporations. In fact, a significant consolidation of the rural electricity distribution and retail business took place in which smaller and financially weak LDUs were absorbed by the larger PCs. In 2010 the Vietnam Distribution Code was approved. It outlined the rights and obligations of PCs and their customers, including provisions regarding quality of service obligations and consumer protection.

Certain features of Vietnam's experience, especially the ways in which the government tackled the challenges that emerged, can provide useful lessons for other countries. The lessons learned are organized in two parts. The first is from the perspective of the government. These lessons can inform the strategies of other developing countries in their efforts to expand access to electricity. The second part came from the World Bank experience. These lessons can be relevant during the development of future projects.

Vietnam's success in rural electrification can be explained by a range of factors:

1. Unwavering government commitment, responsive to strong demand from society
2. Long-term vision, gradual and flexible approach
3. Responsibilities shared by all levels of government, sector participants, and consumers
4. Costs and mobilization of various resources shared by all stakeholders
5. Flexibility in the management and operation of rural networks.

Key lessons learned from Vietnam's experience are summarized below.

Lesson 1: Vietnam's success can be credited to its unwavering national commitment to rural electrification. A significant feature of Vietnam's experience has been the bottom-up manner in which the drive for RE manifested. Demand for electricity was very high across Vietnam. However, one can easily argue that high demand for electricity is present almost everywhere in the world. What was distinctive about the Vietnamese experience was the way that this high demand translated into action and, eventually, results. The local and central governments listened to the people and were responsive to their strong desire for electricity access. Local and central government authorities made RE a priority and mobilized their resources to make it possible. Local authorities' responsiveness to the strong societal demand, their choice to accord adequate priority to this issue, and the culmination of this in a national agenda item were key factors for success. Dedication and collaboration were persistent among central government policymakers; provincial, district, and commune authorities; as well as EVN and local communities.

Once RE targets were set and pledges to support RE were made, policymakers never back-tracked.

Lesson 2: Transformation in the rural sector needs to be seen in a long-term context. The Vietnamese government maintained a strategic vision for achieving its electrification targets. The core tenet was to achieve RE through extending the national grid. An important factor for success was setting targets for a gradual and flexible program based on a realistic analysis of what can be achieved given the available resources.

Lesson 3: The presence of a long-term vision and a systematic plan is important for grid extension. When based on a realistic assessment of what can be achieved within available means, such a plan can help maintain costs at reasonable levels, improve revenues in the early years to support later system expansion, manage expectations, and keep political pressure to extend the grid to unsuitable areas to a minimum. On the other hand, where grid extension to some remote areas is deemed not feasible or economically viable, alternative means to secure access to electricity should be explored. The framework for alternative options including financial and other incentives should be communicated to stakeholders through policy statements and other means.

Lesson 4: Vietnam's electrification success story is rooted in a strategy that was anchored in very clear objectives, implemented gradually, and fine-tuned over time to reflect changing priorities. The central government, in collaboration with various levels of local authorities and stakeholders, continually assessed how to advance the program. Building flexibility into the design and implementation of RE programs also proved useful. All actions were taken without losing sight of the overarching goal of electrification. This unified thought, in turn, contributed to the high-level political goal of national solidarity.

Lesson 5: In the early years of its existence as a commercial entity, Electricity of Vietnam (EVN) stayed out of the rural electricity supply business. However, EVN's emergence as a strong champion for rural electrification in the late 1990s has been an important factor in ensuring the quality and sustainability of rural electricity supply going forward. Despite recognizing the high societal demand and government support for RE, EVN initially was reluctant to engage in rural electricity supply due to concerns over the apparent unprofitability of RE and the limits on its own financing capacity. From 1999 on, the central government made it possible and, even more importantly, profitable, for EVN to participate in RE. Through a series of decisions, such as Decision 22 and Decree 45, the Prime Minister's office equipped EVN with the mandate and resources it needed to lead RE in a financially sustainable way. The resulting reorganization of the rural electricity supply sector had considerable impact on the pursuit of the access agenda.

As Vietnam's experience demonstrates, ensuring the sustainability of rural electricity supply businesses is critical. In the case of expansion of access through large-scale grid extensions, a precondition of sustainability is to secure interest, commitment, and dedication from the country's main utility. These can be obtained by enabling the utility to participate in rural electrification on terms by which it can meet its commercial objectives. The utility should be equipped with a clear mandate and provided with the resources it needs to perform its leadership role in electrifying rural areas through grid extension.

Lesson 6: The presence of cost sharing among different parties has been an important contributor to the success of Vietnam's rural electrification program. In addition to making financing and building rural systems easier, cost sharing helped create a sense of ownership by the parties involved. Financial support by provincial, commune, and district authorities and the Prime Minister's office was a critical element in ensuring the rapid increase in rural access to electricity.

Lesson 7: A well-formulated and properly communicated program with achievable goals and investments from multiple sources has proved very effective. Issuing policy documents that outlined the principles underlying this program was useful in formalizing the government's commitment and sending a clear message to all stakeholders that government resources would be available to backstop local resources.

Lesson 8: The culture of payment is likely to have a significant bearing on the success of rural electrification. It may not be possible to replicate Vietnam's strong culture of payment in other countries. However, countries easily can adopt local involvement in the management and operations of rural electricity networks, particularly bill collection. The service agent model adopted in the early 2000s is a good illustration of how local involvement can be accomplished. This model helped ensure accountability within local communities and minimize nonpayment. It also reduced system losses and lowered PCs' system operation and management costs. Overall, the involvement of local people in managing and operating the LV system played an important role in ensuring the success and sustainability of scaling up electricity access.

Lesson 9: It is important to note that ensuring the financial viability of grid extension for rural electrification does not mean a private-sector style rate of return. Rather, the focus should be on allowing reasonable returns to investors while seeking to make new connections as affordable as possible. When grid extension covers customers who cannot afford to pay the full cost of connections, there should be mechanisms that compensate the investors.

Lesson 10: The separation of responsibilities for MV and LV systems, and the multiplicity of entities being allowed to build, manage, and operate rural networks proved to be very effective in facilitating the rapid expansion of access. However, this rapid build-up of networks was carried out without minimal technical standards for the equipment and materials used. As a result, issues with system performance, including low efficiency, high losses, and low supply quality, led to a series of medium- and long-term costs that eventually had to be recovered.

Thus, it is critical for decisionmakers to be aware of the trade-offs involved in the various options and their long-term consequences. It will be important to carefully weigh the benefits and costs of introducing some kind of minimal technical standards for rural systems and specifications for equipment and materials used.

- If decisionmakers' focus is to rapidly expand access to electricity, the option of allowing multiple entities and financing arrangements for the construction, management, and operation of rural networks[68] is likely to provide the fastest expansion of access.
- Some decisionmakers would like to focus instead on the longer term sustainability of rural electrification. For these decisionmakers, the suitable choice would be to limit the number of actors that can build, manage, and operate rural networks; and to introduce minimal technical standards that should be adhered to. In this case, decisionmakers should expect to move more slowly, as was Vietnam's experience after 2004.

As a general principle, it may be useful to distinguish between sharing costs and sharing responsibilities for building, management, and operation of rural networks. Sharing is a must so it is important to determine the most appropriate ways of combining different funding sources and allocating costs proportionally. On the other hand, it is advisable to allocate responsibilities in a way that will not hamper the coordinated construction, management, and operation of the systems themselves.

Lesson 11: After the initial delay, the beginning of EVN's involvement in the rural energy supply business was a major turning point in ensuring the technical quality of the rural energy networks. In the later years, the introduction of

68. Without imposing minimal technical requirements, as was the case in Vietnam until the late 1990s.

national technical specifications further buttressed improvements in power supply quality, alongside a significant reduction in technical losses, costs, and tariffs. This reduction, in turn, helped generate significant long-term savings.

Lesson 12: Rural electrification is as much about electrifying households as it is about providing electricity for other uses. The initial focus on productive uses may have been helpful to rapidly develop networks. However, to capture the full potential benefits from rural electrification, once basic constraints with respect to availability of supply are overcome, it is crucial not to lose sight of the importance of HH connections.

The Way Forward

Going forward, a number of tasks need to be fulfilled as part of Vietnam's rural electrification program.

1. ***Continued investment in existing distribution networks.*** In the years ahead, significant investment will be needed to rehabilitate existing LV systems to reduce losses, and to upgrade the MV networks to meet growing demand.

   - ***Investment to rehabilitate existing LV systems.*** Many of the small LV systems developed during the 1990s in rural communes remain relatively weak. These systems need to be rehabilitated to reduce losses and to increase the quality and quantity of power supply. A significant amount of resources and effort will need to be dedicated to rehabilitate LV networks in approximately 3,000 communes. Based on preliminary EVN estimates, bringing existing LV systems up to current Vietnamese standards could cost US$2 billion–$3 billion.
   - ***Investment to upgrade MV networks.*** The growth in LV systems and increased demand in rural areas imply greater need to improve the MV distribution network over the medium term. Quality of service is of concern in some rural areas because existing systems are becoming unable to meet existing and projected load requirements. In some instances, the MV systems bottlenecked the power flow from the transmission system to the low voltage systems in the communes.

2. ***Ensuring the sustainability of rural electricity networks.*** In February 2009, Decision 21 made it possible for a PC to take over the management and assets of any local distribution utility (LDU) that was financially weak. The challenge going forward is to ensure that the transfer of the responsibility for managing and operating existing LV systems that came from financially weak LDUs is paced according to the absorptive capacity of the PCs. To be sure, it may be necessary to find the most effective ways to build the institutional capacity of PCs to equip them to manage and operate rural electricity networks efficiently and sustainably.
3. ***Extending access to those still without it.*** Approximately 1 million Vietnamese households, mainly poor and in mountainous areas and islands, still lack access to electricity. The objective of expanding electricity access to unserved rural and mountainous areas of the country was included in the December 2007 National Energy Development Strategy. This strategy envisages that all rural HH will have access to electricity by 2020. The challenge will be to identify the most appropriate way of electrifying these remaining HH.
4. ***Continuing to ensure that electricity is affordable to the poor.*** As the country expands access to a greater number of people, chances are that a greater number of poor people will be connected to the system. The Prime Minister's Decision 21, issued in February 2009, established a clear framework for providing energy to the poor. The decision stipulates an incremental block tariff with a lifeline block that

pays a lower electricity price for the first 50kWh of consumption. Decision 21 also allows the PCs to recover operating costs through the tariffs charged to all consumers in their territory, in effect, cross-subsidizing the remote and poor consumers with revenues generated from other low-cost consumers. If, in the future, policymakers, for any reason, wish to make subsidies more transparent, they will need to make a determination as to how this can be achieved through a new mechanism.

Vietnam needs to make hard choices concerning the future of its rural electrification program. The country's policymakers must determine how to allocate the scarce resources among the country's competing priorities. The lawmakers will have to determine how to allocate resources among investments to rehabilitate LV systems, upgrade MV networks, and expand electricity access to those remaining without it. Simultaneously, policymakers will need to set aside resources to continue to provide for the poor and build institutional capacity for the sustainable management and operation of the rural networks. There is no doubt that improving the existing distribution systems will benefit a larger number of people. On the other hand, completing the last mile of the access agenda to connect the unserved populations is a key priority for the government. The question is how the upgrading can be done sustainably and from what sources the necessary resources will be forthcoming.

APPENDIX 2

Appendix 2. Energy Access Projects Funded in the EAP Region by IBRD, IDA and GEF, 2001–10 *(US$ mil)*

Country	Project	Effective from	Objective	Project cost	IBRD	IDA	GEF
Cambodia	*Rural Electrification and Transmission*	2003	a. Improve power sector efficiency and reliability and reduce electricity supply costs. b. Improve standards of living and foster economic growth in rural areas by expanding rural electricity supplies. c. Strengthen electricity institutions, regulatory framework, and "enabling environment" for sector commercialization and privatization. *Global environment objective*: Overcome barriers to renewable energy development in Cambodia, including those related to lack of policy framework, financing, information, and institutional capacity.	150.0	—	40	5.75
Lao PDR	*Rural Electrification Phase I*	2006	a. Increase access to electricity of rural households in villages of targeted provinces. b. Improve financial performance of power sector.	36.3	10.0	—	5.00
Lao PDR	*Rural Electrification Phase II*	2010		34.8	20.0	—	—
Mongolia	*Renewable Energy for Rural Access*	2006	Increase access to electricity and improve reliability of electricity service among herder population and in off-grid soum centers. REREP addresses 30% of isolated soum centers and 40% of herders' market. *Global development objective:* Remove barriers to development and use of renewable energy technologies in grid- and off-grid-connected systems and reduce CO_2 emissions.	27.0	3.0	—	4.60

(continued)

Appendix 2. Energy Access Projects Funded in the EAP Region by IBRD, IDA and GEF, 2001–10 *(US$ mil) (continued)*

Country	Project	Effective from	Objective	Project cost	IBRD	IDA	GEF
Philippines	*Rural Power*	2004	a. Test and demonstrate viable business models that maximize leveraging of public resources with private investment for decentralized electrification. b. Support transformation of electric cooperatives (ECs) through institutional and operational improvements. c. Avoid CO_2 emissions through wider use of renewable energy.	26.7	10.0	—	9.0
Philippines	*Electrical Cooperative System Loss Reduction*	2004	Achieve significant and sustained energy efficiency improvements in ECs to provide current and prospective viable EC customers with reliable and least-cost power supply over long term. To this end: a. Develop and test financial and contractual mechanisms to support private sector investment, operations and management, and risk-sharing to support system loss reduction measures in selected ECs. b. Support commercial lending to qualified ECs for efficiency improvements	62.3	—	—	12.0
Timor-Leste	*Energy Services Delivery*	2006	Stabilize power services in Dili by restoring or improving operational efficiency, reliability, safety, and availability of power supply; promote sustainability of power sector.	8.5	—	4.5	2.0
Vietnam	*Rural Energy*	2000	a. Expand rural access to electricity in 902 communes located in 34 provinces of Vietnam through extending grid. b. Define and establish institutional mechanisms and strategy for rural electrification. c. Promote application of renewable energy sources in areas inaccessible to national grid and supplement grid power supply.	216.0	—	150.0	—

(continued)

Appendix 2. Energy Access Projects Funded in the EAP Region by IBRD, IDA and GEF, 2001–10 *(US$ mil) (continued)*

Country	Project	Effective from	Objective	Project cost	IBRD	IDA	GEF
Vietnam	*Second Rural Energy*	2005	Improve access to good-quality, affordable electricity services efficiently and sustainably to rural communities to support Vietnam's efforts toward socioeconomic development through: a. Major upgrade and /or expansion of rural power networks in approximately 1,200 communes. b. Conversion of current ad-hoc local electricity management systems to LDUs as legal entities. c. Capacity building assistance for LDUs, provincial authorities, participating regional power companies, and national authorities involved in planning and regulating rural electrification.	329.5	—	200.0	5.25
Vietnam	*Rural Distribution*	2008	Improve reliability and quality of medium voltage service to targeted retail electricity distribution systems.	207.5	—	150.0	—
Vietnam	*Renewable Energy Development*	2009	Increase supply of electricity to national grid from renewable energy sources on a commercially, environmentally, and socially sustainable basis.	318.0	—	202.0	—
Total				**1416.6**	**43**	**746.5**	**43.60**

References

Aden, N., D. Fridley, and N. Zheng. 2009. "China's Coal: Demand, Constraints, and Externalities." Environmental Energy Technologies Division, Lawrence Berkeley National Laboratory, Berkeley, CA.

ADB (Asian Development Bank). 2009. Energy Statistics in Asia and the Pacific Region. 1990–2006. Manila.

Ahmed, K., Y. Awe, D.F. Barnes, M.L. Cropper, and M. Kojima. 2005. "Environmental Health and Traditional Fuel Use in Guatemala." World Bank, Washington, DC.

Aspden, P. 2010. "Museum Delivers on Promise of the Gods." *The Financial Times,* October 14.

Bacon, R., E. Ley, and M. Kojima. 2010. "Subsidies in the Energy Sector: An Overview." Background Paper for the World Bank Group Energy Sector Strategy. World Bank, Washington, DC.

Barnes, D.F. 2007. *The Challenge of Rural Electrification: Strategies for Developing Countries.* Washington, DC: RFF (Resources for the Future) Press.

Barnes, D.F., K. Krutilla, and W.F. Hyde. 2005. *Urban Household Energy Transition: Social and Environmental Impacts in the Developing World.* Washington, DC: Resources for the Future.

Barnes, D.F., S. Khandker, and H. Samad. 2010. "Energy Access, Efficiency, and Poverty: How Many Are Energy Poor in Bangladesh?" Policy Research Working Paper 5332. World Bank, Washington, DC.

Carolina Population Center and National Institute of Nutrition and Food Safety (NINFS) (Beijing). 2008. China Health and Nutrition Survey 2006. University of North Carolina at Chapel Hill. www.cpc.unc.edu/projects/china

Cecelski, E. 1995. "From Rio to Beijing: Engendering the Energy Debate." *Energy Policy* 23 (6): 561–75.

Dherani, M., D. Pope, M. Macharenhas, K. Smith, M. Weber, and N. Bruce. 2008. "Indoor Air Pollution from Unprocessed Solid Fuel Use and Pneumonia Risk in under-5 Children: Systematic Review and Meta-Analysis." *Bulletin of the World Health Organization* 86 (5): 390.

Drouineau, M., N. Mazi, E. Assoumou, and V. Mazauricz. 2010. "TIMES Model for the Reunion Island: Addressing Reliability of Electricity Supply." International Energy Workshop Conference, Stockholm, June 21–23.

Duflo, E., M. Greenstone, and R. Hanna. 2008. "Indoor Air Pollution, Health and Economic Well-Being." Poverty Action Laboratory, MIT (Massachusetts Institute of Technology), Cambridge. www.mit.edu/files/2375

Enis, B., S. Rivera, Z. Boeahmova, and S. Constant. 2006. "Indoor Air Pollution in Cold Climates: The Cases of Mongolia and China." ESMAP (Energy Sector Management Assistance Program) Knowledge Exchange Series 8. World Bank, Washington, DC.

Enis, B., and M. Ezzati. 2007. "Household Energy, Indoor Air Pollution and Health: A Multisectoral Intervention Program in Rural China." ESMAP Special Energy and Poverty Report. World Bank, Washington, DC.

Ezzati, M., H. Saleh, and D.M. Kammen. 2000. "The Contributions of Emissions and Spatial

Microenvironments to Exposure to Indoor Air Pollution from Biomass Combustion in Kenya." *Environmental Health Perspectives* 108: 833–39.

Gilbert, E.M. 2009. "Tax Policies for Low-Carbon Technologies." NBER (National Bureau of Economic Research) Working Paper Series 15054. Cambridge, MA. www.nber.org/papers/w15054

GIZ (Gesellschaft für Internationale Zusammenarbeit). 2011. "Carbon Markets for Improved Cookstoves: A GIZ Guide for Project Operators." HERA (Haushaltenergievorhaben) (Household Energy Services) Report. 4th rev ed. Eschborn, Germany. January.

Golumbeanu, R., and D.F. Barnes. 2010. "A Comparison of Electricity Connection Costs in Developing Countries." Draft paper for Africa Electricity Initiative (AEI), World Bank, Washington, DC.

Government of China. 2008. National Agriculture Census. National Bureau of Statistics of China, Beijing. www.stats.gov.cn/

_____. 2009. Statistics on China Renewable Energy. Science and Education Department, Ministry of Agriculture, Beijing.

IEA (International Energy Agency). 2009. *World Energy Outlook 2009*. Paris: OECD (Organisation for Economic Cooperation and Development)/IEA.

_____. 2010. *World Energy Outlook 2010*. Paris: OECD/IEA.

IEG (Independent Evaluation Group). 2008. "The Welfare Impact of Rural Electrification: A Reassessment of the Costs and Benefits." An IEG Impact Evaluation. Independent Evaluation Group, World Bank, Washington, DC.

IIT (Indian Institute of Technology) and ERI (Energy and Resources Institute). 2010. "New Initiative for Development and Deployment of Improved Cookstoves: Recommended Action Plan." Draft report prepared for the Ministry of New and Renewable Energy, Government of India, New Delhi.

Jia, Z. 2011. "Summary Review of the Cooking Stove Program in China." Background Paper prepared for the East Asia and the Pacific Infrastructure Unit, World Bank, Beijing.

Kammen, D.M. 1995. "Cookstoves for the Developing World." *Scientific American* 273 (1) (July): 72–75.

Kammen, D.M., R. Bailis, and A.V. Herzog. 2002. "Clean Energy for Development and Economic Growth: Biomass and Other Renewable Energy Options to Meet Energy and Development Needs in Poor Nations." UNDP (United Nations Development Programme) and Government of Morocco, New York.

Lighting Africa. 2011. www.Lightingafrica.org

MacCarty, N., D. Ogle, D. Still, T. Bond, C. Roden, and B. Wilson. 2007. "Laboratory Comparison of the Global-Warming Potential of Six Categories of Biomass Cookstoves." Aprovecho Research Center, Creswell, OR.

MacCarty, N., D. Still, and D. Ogle. 2010. "Fuel Use and Emissions Performance of Fifty Cookstoves in the Laboratory and Related Benchmarks of Performance." *Energy for Sustainable Development* 14 (3) (September): 161–71.

Masera, O.R., B.D. Saatkamp, and D.M. Kammen. 2000. "From Fuel Switching to Multiple Cooking Fuels: A Critique of the Energy Ladder Model in Rural Households." *World Development* 28 (12): 2083–103.

MIME (Ministry of Industry, Mines and Energy). 2009. "Cambodia Rural Electrification Strategy and Implementation Plan (CRESIP)." Royal Government of Cambodia, Phnom Penh.

Mumssen, Y., L. Johannes, and G. Kumar. 2010. "Output-Based Aid: Lessons Learned and Best Practices." GPOBA (Global Partnership on Output-Based Aid), World Bank, Washington, DC.

Niez, A. 2010. "Comparative Study on Rural Electrification Policies in Emerging Economies: Keys to Successful Policies." Information Paper. IEA (International Energy Agency), Paris.

Parikh J., K. Balakrishnan, V. Laxmi, and H. Biswas. 2001. "Exposure from Cooking with Biofuels: Pollution Monitoring and Analysis for Rural Tamil Nadu, India." *Energy* 26: 949–62.

PPI (Pacific Policy Institute). 2009. "Social and Economic Impacts of Introducing Telecommunications in Vanuatu." Port Vial, Vanuatu.

REN21 (Renewable Energy Policy Network for the 21st Century). 2010. "Renewables 2010 Global Status Report." REN21 Secretariat, Paris.

Republic of the Philippines. 2009. "Missionary Electrification Development Plan for 2009–2013." Electric Power Industry Management Bureau, Department of Energy, Manila.

Savage, P. 2010. "Information Note on DC Microgrids." Nextek Power Systems, Inc., Detroit, MI.

Sinton, J., R. Kirk, J.W. Peabody, L. Yaping, Z. Xiliang, R. Edwards, and G. Quan. 2004. "An Assessment of Programs to Promote Improved Household Stoves in China." *Energy for Sustainable Development* VIII (3) (September).

Smith, K.R., S. Gu, K. Huang, and D. Qiu. 1993. "100 Million Improved Cookstoves in China: How Was It Done?" *World Development* 21 (6) (June): 941–61.

Smith, K.R., S. Mehta, and M. Maeusezahl-Feuz. 2004. "Indoor Smoke from Household Solid Fuels." In *Comparative Quantification of Health Risks: Global and Regional Burden of Disease Attributable to Selected Major Risk Factors*, ed. M. Ezzati, A.D. Rodgers, A.D. Lopez, and C.J.L. Murray. Geneva: World Health Organization.

Smith, K.R., and K. Deng. 2010. "A Chinese National Improved Stove Program for the 21st Century to Promote Rural Social and Economic Development." *Energy Policy Research* 1: 23–35.

SPC (Secretariat of the Pacific Community). 2010. "Toward an Energy-Secure Pacific: A Framework for Action on Energy Security in the Pacific (2010–20)." Noumea, New Caledonia.

SPREP (Secretariat of the Pacific Regional Energy Programme). 2004. Pacific Regional Energy Assessment, Pacific Island Renewable Energy Project (PIREP), Regional Overview. Apia, Samoa.

Tian, C. 2002. "Rapid Expanding of China LPG Market." *International Petroleum Economics* 10 (5).

Tuntivate, V. 2010. "Promoting the Uses of LPG for Cooking in Thailand." Paper prepared for Energy Transport and Water Department, World Bank, Washington, DC.

UN (United Nations) Foundation. 2010. "Global Alliance for Clean Cookstoves." Fact Sheet. www.unfoundation.org/assets/pdf/global-alliance-for-clean-cookstoves-factsheet.pdf

UN-DESA (United Nations Department of Economic and Social Affairs). 2008. *Demographic Yearbook 2008*. New York.

UNEP (United Nations Environment Programme) Risoe Centre. 2010. "CDM/JI (Clean Development Mechanism/Joint Implementation) Pipeline Analysis and Database," Frederiksborgvej, Denmark. www.cdmpipeline.org

UNFCC (United Nations Framework Convention on Climate Change). 2010. "CDM Project Co-benefits in Hubei Province, China: Biogas Digesters Enhance the Welfare of Low-Income Rural Communities." Fact Sheet prepared for United Nations Framework Convention on Climate Change, New York. http://cdm.unfccc.int/about/ccb/CDM_Cobenefits_Hubei_China.pdf

von Moltke, A., C. McKee, and T. Morgan. 2003. *Energy Subsidies: Lessons Learned in Assessing Their Impact and Designing Policy Reforms*. Sheffield, England: Greenleaf Publishing for UNEP.

Wang, X., N. Berrah, S. Mathur, and F. Vinuya. 2010. *Winds of Change: East Asia's Sustainable Energy Future*. Washington, DC: World Bank.

WHO (World Health Organization). 2005. "WHO Air Quality Guidelines Global Update 2005." Copenhagen.

______. 2006. "Fuel for Life: Household Energy and Health." Geneva.

______. 2009. "Country Profiles of Environmental Burden of Disease." Geneva.

WHO and UNDP. 2009. *The Energy Access Situation in Developing Countries: A Review Focusing on the Least Developed Countries and Sub-Saharan Africa*. New York: UNDP.

World Bank. 1996. "Rural Energy and Development: Improving Energy Supplies for Two

Billion People." Development in Practice Series. Washington, DC. September.

_____. 2002. "India: Household Energy, Indoor Air Pollution and Health." ESMAP Report 261/02. Washington, DC.

_____. 2005. "Rural Electrification in Tunisia: National Commitment, Efficient Implementation and Sound Finances." ESMAP Report 307/05. Washington, DC.

_____. 2006a. "Indoor Air Pollution in Cold Climates: The Cases of Mongolia and China." ESMAP Knowledge Exchange Series 8. Washington, DC.

_____. 2006b. "Sub-Saharan Africa: Low-Cost Methods in Electricity Distribution Networks." ESMAP Working Paper 104/06. Washington, DC.

_____. 2007a. "Household Energy, Indoor Air Pollution and Health: A Multisectoral Intervention Program in Rural China." ESMAP Special Energy and Poverty Report. Washington, DC.

_____. 2007b. "Technical and Economic Assessment of Off-Grid, Minigrid and Grid Electrification Technologies." ESMAP Technical Paper 121/07. Washington, DC.

_____. 2008a. "Argentina–Renewable Energy in the Rural Market Project: Additional Financing." Project Paper. Washington, DC.

_____. 2008b. "Electricity beyond the Grid: Innovative Programs in Bangladesh and Sri Lanka." ESMAP Knowledge Exchange Series 10. Washington, DC.

_____. 2008c. Eco-Farming Project. Project Appraisal Document. Report N. 39781-CN. Washington, DC.

_____. 2008d. "Tackling Rural Electrification: A Multi-Track Approach." ESMAP Background Paper. Washington, DC.

_____. 2008e. *World Development Indicators 2008*. Washington, DC.

_____. 2009. "Uganda: Energy for Rural Transformation Program." Implementation Completion Report. ICR1288. Washington, DC.

_____. 2010a. "Addressing the Electricity Access Gap." Background Paper to the World Bank Group's Energy Strategy. Washington, DC.

_____. 2010b. "Environmental Pilot Program within the Government of Indonesia's National Program for Community Empowerment in Rural Areas (PNPM-Rural) Status Report. Indonesia Social Development Sector, Jakarta.

_____. 2010c. "Improved Cookstoves and Better Health in Bangladesh: Lessons from Household Energy and Sanitation Programs." ESMAP Report. Washington, DC.

_____. 2010d. "Rural Poverty Reduction in Northeast Brazil: Achieving Results through Community-Driven Development." Washington, DC.

_____. 2010e. *World Development Indicators*. Washington, DC.

_____. 2011a. "Household Cookstoves, Environment, Health, and Climate Change: A New Look at an Old Problem." Paper prepared for the Climate Change Unit of the Environment Department, Washington, DC.

_____. 2011b. "Vietnam's Rural Electrification Story: State and People, Central and Local, Working Together." ASTAE (Asia Sustainable and Alternative Energy Program), Washington, DC.

_____. 2011c. "Wood-Based Biomass Energy Development for Sub-Saharan Africa: Issues and Approaches." Africa Energy Unit, Washington, DC.

_____. Forthcoming 2011. "Power to the People: Twenty Years of National Electrification Effort in Lao PDR." East Asia Sustainable Development Department, Washington, DC.

Helpful URLs

Selected national electricity utility/institution websites:

Electricité du Cambodge (EDC)	www.edc.com.kh
Electricité du Laos (EDL)	www.edl-laos.com
Electricity of Vietnam (EVN)	www.evn.com.vn
Fiji Electricity Authority (FEA)	www.fea.com.fj
National Electricity Authority (NEA) (Philippines)	www.nea.gov.ph
Perusahaan Listrik Negara (PLN) (Indonesia)	www.pln.co.id
PNG Power Ltd	www.pngpower.com.pg
Provincial Electricity Authority (PEA) (Thailand)	www.pea.co.th

Other relevant URLs

Community Development Carbon Fund	www.go.worldbank.org/QNLHGWLPS0
Global Alliance for Clean Cookstoves	http://cleancookstoves.org
Groupe Energies Renouvelables, Environnement et Solidarités	www.geres.eu
Global Alliance for Clean Cookstoves	http://cleancookstoves.org/
Lighting Africa Project	www.lightingafrica.org
Water and Sanitation Program	www.wsp.org

ECO-AUDIT

Environmental Benefits Statement

The World Bank is committed to preserving endangered forests and natural resources. The Office of the Publisher has chosen to print ***One Goal, Two Paths*** on recycled paper with 30 percent post-consumer waste, in accordance with the recommended standards for paper usage set by the Green Press Initiative, a non-profit program supporting publishers in using fiber that is not sourced from endangered forests. For more information, visit www.greenpressinitiative.org.

Saved:
- 14 trees
- 6 million BTU of total energy
- 1,434 lbs. of CO_2 equivalent of greenhouse gases
- 6,468 gallons of wastewater
- 410 lbs. of solid waste

green press INITIATIVE